T0351031

Real World AI Ethics for Data Scientists

In the midst of the fourth industrial revolution, big data is weighed in gold, placing enormous power in the hands of data scientists – the modern AI alchemists. But great power comes with greater responsibility. This book seeks to shape, in a practical, diverse, and inclusive way, the ethical compass of those entrusted with big data.

Being practical, this book provides seven real-world case studies dealing with big data abuse. These cases span a range of topics from the statistical manipulation of research in the Cornell food lab through the Facebook user data abuse done by Cambridge Analytica to the abuse of farm animals by AI in a chapter co-authored by renowned philosophers Peter Singer and Yip Fai Tse. Diverse and inclusive, given the global nature of this revolution, this book provides case-by-case commentary on the cases by scholars representing non-Western ethical approaches (Buddhist, Jewish, Indigenous, and African) as well as Western approaches (consequentialism, deontology, and virtue).

We hope this book will be a lighthouse for those debating ethical dilemmas in this challenging and ever-evolving field.

Chapman & Hall/CRC Data Science Series

Reflecting the interdisciplinary nature of the field, this book series brings together researchers, practitioners, and instructors from statistics, computer science, machine learning, and analytics. The series will publish cutting-edge research, industry applications, and textbooks in data science.

The inclusion of concrete examples, applications, and methods is highly encouraged. The scope of the series includes titles in the areas of machine learning, pattern recognition, predictive analytics, business analytics, Big Data, visualization, programming, software, learning analytics, data wrangling, interactive graphics, and reproducible research.

Data Science: An Introduction
Tiffany-Anne Timbers, Trevor Campbell, and Melissa Lee

Tree-Based Methods: A Practical Introduction with Applications in R
Brandon M. Greenwell

Urban Informatics: Using Big Data to Understand and Serve Communities
Daniel T. O'Brien

Introduction to Environmental Data Science
Jerry Douglas Davis

Cybersecurity Analytics
Rakesh M. Verma and David J. Marchette

Exploratory Model Analysis: Explore, Explain, and Examine Predictive Models
Przemyslaw Biecek and Tomasz Burzykowski

For more information about this series, please visit: https://www.routledge.com/Chapman–HallCRC-Data-Science-Series/book-series/CHDSS

Real World AI Ethics for Data Scientists
Practical Case Studies

Nachshon (Sean) Goltz
and Tracey Dowdeswell

**In Collaboration with Peter Singer
and Yip Fai Tse**

CRC Press
Taylor & Francis Group
Boca Raton London New York

CRC Press is an imprint of the
Taylor & Francis Group, an **informa** business
A CHAPMAN & HALL BOOK

Designed cover image: © Getty Images

First edition published 2023
by CRC Press
6000 Broken Sound Parkway NW, Suite 300, Boca Raton, FL 33487-2742

and by CRC Press
4 Park Square, Milton Park, Abingdon, Oxon, OX14 4RN

CRC Press is an imprint of Taylor & Francis Group, LLC

© 2023 selection and editorial matter, Nachshon (Sean) Goltz and Tracey Dowdeswell; individual chapters, the contributors

Reasonable efforts have been made to publish reliable data and information, but the author and publisher cannot assume responsibility for the validity of all materials or the consequences of their use. The authors and publishers have attempted to trace the copyright holders of all material reproduced in this publication and apologize to copyright holders if permission to publish in this form has not been obtained. If any copyright material has not been acknowledged please write and let us know so we may rectify in any future reprint.

Except as permitted under U.S. Copyright Law, no part of this book may be reprinted, reproduced, transmitted, or utilized in any form by any electronic, mechanical, or other means, now known or hereafter invented, including photocopying, microfilming, and recording, or in any information storage or retrieval system, without written permission from the publishers.

For permission to photocopy or use material electronically from this work, access www.copyright.com or contact the Copyright Clearance Center, Inc. (CCC), 222 Rosewood Drive, Danvers, MA 01923, 978-750-8400. For works that are not available on CCC please contact mpkbookspermissions@tandf.co.uk

Trademark notice: Product or corporate names may be trademarks or registered trademarks and are used only for identification and explanation without intent to infringe.

ISBN: 978-1-032-27506-2 (hbk)
ISBN: 978-1-032-27505-5 (pbk)
ISBN: 978-1-003-29312-5 (ebk)

DOI: 10.1201/9781003293125

Typeset in Bembo
by KnowledgeWorks Global Ltd.

Contents

Acknowledgement

We would like to thank the following contributors for their excellent and diverse contributions to this book: Peter Hershock (Buddhist ethics), John Hacker-Wright (Virtue ethics), Samuel J. Levine and Daniel Sinclair (Jewish ethics), Colin Marshall (Deontological ethics), Joey Miller and Andrea Sullivan-Clarke (Indigenous ethics), and John Murungi (African ethics). Our aim is to draw an ethical picture that would be as diverse as possible and go beyond the traditional Western approaches to ethics. It is only with the contributions of these esteemed scholars, that we were able to create this book.

Collaborators and Contributors

Collaborators

Peter Singer
University Center for Human
    Values
Princeton University

Yip Fai Tse
University Center for Human
    Values
Princeton University

Contributors

John Hacker-Wright
Department of Philosophy
University of Guelph

Peter D. Hershock
East-West Center

Samuel J. Levine
Touro Law Center

Colin Marshall
Department of Philosophy
University of Washington

Joey Miller
Department of Philosophy
West Chester University

John Murungi
Department of Philosophy
Towson University

Daniel Sinclair
Fordham University Law School

Andrea Sullivan-Clarke
Philosophy Department
University of Windsor

1 Introduction

Moral Machines

Mathematical science shows what is. It is the language of unseen relations between things. But to use and apply that language, we must be able to fully appreciate, to feel, to seize the unseen, the unconscious.

~ (Ada Lovelace 1815–1852)

Introduction: Moral Machines

This book is a manual of ethics for data scientists – and also for lay persons interested in the important ethical issues that arise in data science. Ethics is broadly understood as "a set of moral principles: a theory or system of moral values."[1] It is about how we choose between right and wrong. This book is intended to guide data scientists in applying ethical principles to moral dilemmas involving data. Moreover, it is intended to shape data scientists' perception of right and wrong through guided self-reflection, so as to prevent unethical use of data.

The public's knowledge of data science and machine learning (ML) can seem to be inversely proportional to its importance in our daily lives. This leaves ordinary people quite powerless in understanding – let alone in shaping or governing – these powerful technologies. At the same time, many scientists working in this field have not had either the opportunity to learn about ethics or sufficient resources to turn their minds to the ethical and social impacts of the technologies they are developing. This book is written in the hope of mitigating these deficits.

When one of the authors was a young student in the hard sciences, lectures and workshops on ethics were just starting to become common. Students would sometimes turn up their noses at the idea that scientists should be learning ethics. They would make comments to the effect that, "you can't teach anyone anything about ethics. It's just made up by people" – unlike the scientific method, which had apparently been revealed to us from a higher source. Part of this statement is true: ethics is in fact made up by people. This is one of its chief advantages – it is always and ever people who interpret and apply these values, rules, and precepts, often under dynamic and rapidly evolving social conditions. The choices we make are an important part of how we

DOI: 10.1201/9781003293125-1

adapt to and live ethically in the world. Today, we face a revolution in the powerful, at times wondrous – and often highly risky – field of data science. This requires a new thinking and a close engagement with the ethics of ML technologies. Ethical issues arising from these technologies can, and should, be taught widely.

To render the manual as friendly and useful as possible, we have built the book around seven case studies involving the misuse of data. We then reached out to leading ethics scholars to comment on the ethical dilemmas each case presents. In order to present humanity's wide spectrum of ethical approaches, we went beyond the three Western approaches to ethics – consequentialism (utilitarianism), deontology, and virtue ethics – and have been fortunate to be able to include discussions on Buddhist, Jewish, Indigenous, and African ethics as well.

Each one of these approaches could be used to examine and better understand each of the case studies, as well as the myriad of situations and ethical dilemmas that we have not had the opportunity to present here – and the many more that we cannot yet imagine and that will inevitably arise. These are not the only approaches relevant to understanding data science.[2] They are widespread and influential ethical approaches, and they are used and held in high esteem by a great number of the world's people. Our intention is to provide the reader with a thorough understanding of each approach – a tool kit for future challenges you will encounter in your work with data. While a specific dilemma could be analysed differently by different ethical approaches, the ability to think and analyse the dilemma from different perspectives is a necessary part of the critical thinking process. We hope this book will provide the reader with a broader and more nuanced view of the issues.

What Is Data Science?

In this book, we and the commentators refer at times to artificial intelligence (AI),[3] ML,[4] data science, and computer-assisted decision-making. The case studies and commentaries described in this book are intended to apply to each of these, overlapping, fields.

Data science is broadly "a set of principles, problem definitions, algorithms, and processes for extracting non-obvious and useful patterns from large data sets."[5] AI, on the other hand, can be defined generally as any type of artificial information-processing that carries out a psychological function – such as "perception, association, prediction, planning, motor control" – that up until now has been performed only by living beings. There have traditionally been two basic approaches to developing AI: Good Old-Fashioned AI (GOFAI) and ML. GOFAI, sometimes also called symbolic AI, takes a top-down view of intelligence.[6] The central focus of GOFAI is "humans' ability to understand the world by forming symbolic representations"; these symbols are then developed into rules, which are then "formalized in a way that captures everyday knowledge."[7]

ML is a subset of data science[8] and a large and growing part of the field. Unlike GOFAI, ML is a form of AI that uses a statistical, rather than symbolic, approach to find patterns in a messy world of ambiguity.[9] ML was in many ways a response to the early failures of symbolic AI, which quickly broke down outside of the controlled laboratory environments – unable to process the complexities of real-world situations.[10] ML emphasises prediction – as opposed to the field of statistics, which emphasises associations and explanations.[11]

ML algorithms are trained with the given successive layers of information, presented one at a time, and they can learn in ways that are relatively unsupervised, and relatively mysterious – much more like how our human brains learn.[12] Supervised learning involves algorithms learning to use a function and applying it to a set of data – such as a function describing 'spam' being applied to a set of e-mails to filter out which ones are likely to be spam. The attributes of 'spam' and 'not spam' will be labelled for the algorithm by human beings, who already know the difference.[13] Unsupervised learning, on the other hand, involves no labelling – indeed, we may not even know what qualities we are looking for. The algorithm first examines a set of data and figures out what the relevant attributes are before it applies them to new data.[14] An unsupervised algorithm might, for example, look at numerous images of dogs and determine what set of attributes gives rise to the essence of 'dogness.' When presented with any new picture, it will then decide if it is or is not a dog.

Data science tools have become more user-friendly, and this has opened the field to many new entrants – even those with little training. This means that "it has never been easier to do data science badly," and the unethical consequences of badly designed or executed projects can and should be anticipated by those whose task is to design and deploy those systems.[15]

As Kelleher explains, *data* is an element that has been abstracted from the real world; *information* is data that has been organized to make it understandable and useful; *knowledge* is an accurate understanding of the information the data is giving to us (no easy task); but the most valuable of all is *wisdom*, which we attain when we apply our knowledge in a way that leads to good decision-making.[16] We aim to help our readers develop this kind of wisdom, which we think is at the heart of data science ethics. What are the harmful versus beneficial effects of a given system? Does it take account of the broader social and value systems within which it operates? Is it fair? – including those who are the worst off and who will be most negatively affected?

Ethics, then, is just a part of doing data science well. This means understanding a given real-world problem as being far more than its technical aspects, and also how the finished system will impact the lives and interests of those who are going to be affected by this system.[17]

The Case Studies

Case 1 – Research Ethics and the Scientific Method

The first case study examines research ethics and the foundations of the scientific method used by data scientists in designing ML processes. This case study introduces the reader to such concepts as reproducibility, accuracy, and validity. Much of this discussion has been informed by recent efforts in psychology, as well as the social and health sciences, to tackle the fact that a significant proportion of all published results cannot be reproduced or validated.

This case study focuses on the findings of research misconduct at the Cornell Food & Brand Lab led by Brian Wansink. He was found to have been engaged in several unethical and scientifically invalid research practices, including cherry-picking (making public only positive results), HARKing (hypothesizing after experimental results are known), and p-hacking (manipulating data to obtain a statistically significant result). At the end of the day, good epistemology is also good ethics.

Singer and Fai Tse provide a commentary on Wansink's conduct from the point of view of utilitarianism. They discuss the Ethical Altruism movement, and the importance of using empirical and scientifically sound evidence in moral decision-making. Those who have taken it upon themselves to produce the empirical evidence that others rely on have a weighty obligation indeed. They must do so in a way that is in line with the very best of the scientific method.

Case 2 – Machine Models in Court

ML algorithms have been used as evidence against defendants in a growing number of criminal cases, and hence have raised several ethical issues. Even well-validated models will perform differently in different social contexts. Even excellent models will be deployed by human beings and can give rise to unintended consequences – including bias, prejudice, and deliberate misuse. The use of Markov chain Monte Carlo (MCMC) models also raises the difficult – and unresolved – ethical question of how we assess the evidence produced by these models when they are, by their very nature, not reproducible.

These issues are illustrated through a study of DNA mixture algorithms and their role in the wrongful prosecution of Oral Nicholas Hillary for the murder of a young boy in Potsdam, New York. The police investigation and prosecution contained very strong elements of personal and racial bias against Hillary – a popular and successful soccer coach of Afro-Caribbean heritage. This resulted in a very biased interpretation of the DNA evidence produced to implicate Hillary. The evidence was found to be unreliable and was correctly excluded by the court, resulting in Hillary's acquittal.

Samuel J. Levine examines the use of ML models in the criminal justice system from the point of view of Jewish ethics, and its long-standing engagement with the tension – indeed the paradox – between determinism and free will. We are all moral agents responsible for our choices. But if our actions are foreordained, then are we not judging people for decisions they had no choice but to make? This can be seen, for example, in the widespread use of machine models to predict rates of recidivism and to fashion sentences and decide if accused persons should receive bail.[18] Much of our criminal law is based upon ideas about free will derived from Judaism and promulgated through the Abrahamic religions. The tensions between determinism and free will permeate much of the decision-making in our criminal justice system. Meanwhile, those in pursuit of power rather than justice can misuse scientific technologies in ways that transgress the bounds of morality.

Colin Marshall looks at the evidence produced by machine models from the perspective of deontological ethics, which has long been concerned about identifying and ruling out problematic forms of partiality in ethical decision-making that advantage some people over others. Ethical decisions must pass the universalization test: an action is universal, rather than partial, if all agents in a similar situation would make a similar choice. This should encourage all data scientists to look at the effects of their models from the point of view of those who will be affected. This requires that ethical agents step out of their own, sometimes narrow and biased, point of view. What if everyone behaved this way? What kind of system would we want if we ought to be wrongfully accused by a partial analyst, or a biased algorithm? Systems that include illegitimate partiality are ethically impermissible and should not be used.

Case 3 – Synthetic Media and Political Violence

This case study examines two recent controversies over 'synthetic media.'[19] The first is a 2019 video purporting to be a New Year's Eve address by Gabonese President Ali Bongo. The video was designed to quell fears over Bongo's recent illness, but it inflamed political tensions when it was widely denounced as a 'deepfake.' On January 7, 2019, soldiers of the Republican Guard launched a failed *coup d'état* in Libreville on the grounds that Bongo was no longer in charge, and the ruling party could not be trusted. The coup was put down violently, resulting in the deaths of two Guards and the detentions of many more. It later turned out that the video was genuine after all. Bongo was recovering from a stroke, and it was the effects of his neurological condition – and not manipulated media – that produced the effects seen in the video.

The second example looks at 'shallowfakes' deployed against women in India, particularly journalists and politicians who criticise the ruling party.

These manipulated media include fake auction sites purporting to 'sell' the women, and which depict them in sexually degrading circumstances.

These two examples raise some difficult ethical issues with synthetic media. The political violence in Gabon shows that problems arise not simply because people cannot tell whether the media has been manipulated, but because people no longer trust key social institutions at all. Female journalists in India have found that shallowfake pornography can be soul-destroying and career-ending simply because it has been made public, and not because it is believed by anyone to be 'true.'

Indeed, synthetic media is entering our lives in more and more contexts. Content that we read is increasingly produced by AI – a phenomenon that has recently cropped up even in peer-reviewed scientific journals.[20] What does it mean to live in a world of manipulated media – one in which truth can no longer be reliably determined and agreed upon, or even one in which truth no longer matters? AI algorithms may help us to identify and remove synthetic media, but they cannot fix these deeper problems.

Singer and Fai Tse look at the problems posed by synthetic media through the lens of utilitarian ethics. They emphasise the importance of truth, which means using the scientific method and empirically sound evidence to make decisions. To do otherwise only reinforces the widespread lack of trust over our key institutions. Trust and strong institutions that promote the well-being of those in our society are built by our valuing *truth*.

Deep- and shallowfake pornography is particularly corrosive of well-being and is rejected by utilitarian ethics. It serves to reinforce the view that women are to be used as objects for the entertainment of others. This can cause many different harms to women, including intimidation, oppression, and forcing women out of roles in public life.

Murungi writes from the perspective of Ubuntu ethics. He encourages data scientists to appreciate their role as architects of the data-driven world we live in and to reflect on the ramifications of their world-creation. He notes that data science is currently taught as a neutral, apolitical endeavour, yet its effects on African peoples are anything but neutral. In Indigenous African ethics, the well-being of the community is both ethical and political, incorporating both the individual and the community. AI undermines both, by trespassing against societies in Africa and around the world, and by undermining our communal sense of 'we' – of the Indigenous social order and social values that form the basis of authentic and ethical lives. Murungi states the inevitable, yet ominous, fact that the "difficulty of distinguishing what is real from what is fake will continue to haunt us."

Miller and Sullivan-Clarke use Indigenous ethics to discuss the many ways how data can be weaponised to manipulate, coerce, control, oppress, and disenfranchise certain groups. Indigenous peoples have often been targets of weaponised data in this way. This has led to the burgeoning Indigenous Data Sovereignty Movement that seeks to give Indigenous peoples back their autonomy and control over their data and how it is used – and from this movement, we can all learn to chart a path to freedom and autonomy in a world of weaponised data.

Case 4 – Biometrics and Facial Recognition

This case study examines the ethical issues arising from the use of biometrics as a form of identification. It is vitally important in a technologically developed world that individuals have a valid means of authenticating their identity and preventing access to their private data. Biometric forms of identification solve many problems with ID: they are individualizing, they are accurate and reliable, and they are more secure than other forms of ID. However, biometric ID systems can give rise to many ethical problems when they are used by authorities for identification, forensics, and surveillance. Here, we look at the unauthorized use of facial recognition by the Royal Canadian Mounted Police (RCMP), which was found to be a violation of Canadian privacy laws by the Privacy Commissioner.

The RCMP had used a system produced by a company named Clearview AI to search for suspects and to locate child victims of online sexual exploitation. The photographs used by Clearview had been taken from social media and other internet sites and were nominally 'public.' The RCMP therefore argued that their use of the systems and these photos did not violate privacy laws. However, even data taken from public websites can give rise to privacy issues when users do not know and have not consented to, novel and intrusive uses of their data.

Dowdeswell and Goltz comment that Clearview AI's decision to roll out its facial recognition technology in the war in Ukraine violates both the international laws of armed conflict as well as the values of humanitarianism that gave rise to them. We argue that data systems that have the potential to target civilians, or to otherwise violate the laws of armed conflict, are unethical and their use should be prohibited.

Miller and Sullivan-Clarke analyse biometric data from the perspective of Indigenous values. Biometric data is not unethical in and of itself, but it might be used in ways that are immoral and harmful. This is well appreciated by Indigenous persons, who have often experienced data being used against them in ways that are oppressive and that justify a militarized response. We need to move beyond a narrow, Western, framing of the issue that prioritizes individual privacy over other values. We can use Indigenous values to ask whether the use of biometric data will disrupt balance and harmony in social relations, or whether it will be used to harm marginalized communities. Will the use of biometric data in criminal proceedings gives rise to a sense of overconfidence in the accused's guilt? – and thus discourage humility, an important Indigenous value. The project of truth and reconciliation requires that Indigenous peoples and groups be consulted about the uses of their data and that data governance promotes the sovereignty and autonomy of Indigenous peoples.

Case 5 – Content Moderation: Dangerous Speech and Ethnic Cleansing in Myanmar

This case study examines the use of AI in content moderation, using the example of how anti-Rohingya speech posted on Facebook has been linked to their ethnic cleansing by government forces in Myanmar. This issue has generated some very contentious discussions concerning what content should be prohibited on social media platforms and how takedown processes should operate. ML algorithms must work in tandem with human content moderators to be effective. At the same time, human content moderation has proven to be a dangerous activity, with several lawsuits filed by content moderators against social media companies for trauma and mental health deterioration that they experienced due to the rigors of viewing harmful content.

Social media companies are in the process of developing content moderation guidelines, and it is not always clear what speech should be prohibited. Prior restraints on the speech of the type being identified and removed by AI content moderation systems could have a chilling effect on free expression and the free exchange of ideas. For those who have had content removed, they are given little notice regarding the takedowns or the reasons why, and there is often no recourse for those whose speech has been restrained. At the same time, those who have been harmed by content on social media – through threats, harassment, fake porn, radicalization, or doxing – often have few remedies against social media platforms.

Hershock analyses content moderation from the perspective of Buddhist ethics. Facebook has a moral responsibility to ensure that its pursuit of commercial self-interest does not cause harm, and it shirked these responsibilities when it permitted hate speech against Rohingya to circulate on its platform. Hershock notes the blurring of the ethical boundaries by platforms like Facebook. This diffuses responsibility – as well as harms – among a diverse array of organizations, individuals, and agents.

In deciding how content should be moderated in the future, Buddhist values have much to teach us about right and corrective speech – which rejects the abuse, gossip, backbiting, slander, lies, and hatred that are so prevalent on social media. It is the relational qualities valued in Buddhism – "compassion, lovingkindness, equanimity, and joy in the good fortune of others" – that are so obviously lacking and that we can cultivate in *all* of our interactions with others, including on social media.

Hacker-Wright looks at hate speech and content moderation from the perspective of virtue ethics. What kinds of virtues are to be promoted – or side-lined – through our responses to content moderation? We can never remove all harmful content from social media, nor will this alone promote a flourishing society. But we can each develop the kind of virtues that will inoculate us against the worst harms of manipulative media. After all, manipulative media is effective precisely because we allow it to reinforce our own pre-existing views. In this way, we are all unwitting – and sometimes willing – participants in social media manipulation. We can counteract this by cultivating such important virtues as moral courage, critical thinking, and a desire to promote the well-being of all those in our global community (including those who disagree with us). In this way, we can achieve the kind of practical wisdom that arises out of the habitual and mindful cultivation of the virtues that Aristotle believed would lead to well-being.

Miller and Sullivan-Clarke look at content moderation through the lens of Indigenous ethics, which views all things as being interconnected. Facebook failed to appreciate how the content on their platform would affect the Rohingya. Their algorithms prioritized highly engaging speech, even when it promoted hatred and fostered violence against a vulnerable minority group. These actions engendered the kind of disharmony and imbalance that so often causes harm. Facebook also failed to cultivate the important value of intellectual humility. Those who design and deploy ML systems have a responsibility to be aware of the limitations of their algorithms as well as their potential for misuse. Miller and Sullivan-Clarke make the important point that several contributors acknowledge: that *words have power*. Algorithms that promote or obscure certain speech over others also have power, and they must be wielded with both humility and a due consideration for the well-being of others.

Case 6 – Mental Malware: Algorithms and Choice Architecture

In 2013, data analytics company Cambridge Analytica began harvesting information on Facebook to develop in-depth psychological profiles on tens of millions of users without their consent. This data was then sold to marketers, including several political campaigns. The scandal resulted in the bankruptcy of Cambridge Analytica, and billions of dollars in fines for Facebook.

The Cambridge Analytica scandal showed how easy it is to harvest sensitive psychological information on users of social media and to use this data in ways that manipulate them – often against their interests. This is a high-profile example of what we term 'mental malware.' Mental malware is frequently deployed against users in ways that are designed not merely to *predict* their behaviour, but to *change* their behaviour – to 'nudge,' manipulate, and alter both individual behaviours and public opinion.

It is easy for political actors to take advantage of the psychological power of algorithms. Since the Cambridge Analytica scandal, further criticisms have been made that social media companies are failing to tackle false and manipulated 'likes,' 'followers,' and other forms of synthetic and inauthentic engagement and that this has been used to manipulate elections and crack down on political dissidents in countries such as Honduras and Azerbaijan.

Hershock looks at content moderation from the perspective of Buddhist ethics. All platform companies filter and moderate content for their users all the time, and we need to be mindful of how they are making these choices, who is choosing the values that guide this process, and what values are being prioritized. The digital choice architecture we establish – and we do have a choice to supersede the platform companies and design our own architecture – should enhance personal and social well-being. While there is a need to foster personal freedom, we must be aware that this has the potential to lock users into their own past choices, thus restricting their freedom. This might simply result in masses of humanity "leading lives in which it is never necessary to learn from mistakes or engage in adaptive conduct." In Buddhist ethics, all things are interconnected. The interrelated infrastructure of digital consciousness we are creating will affect not only our choices, our conduct, and our social relations, but it will fundamentally alter what we *are*.

Hacker-Wright continues his discussion on making authentic choices by cultivating virtues. Aristotle described the virtuous agent as one who pursues the good and avoids vice. A continent agent will aim after the good, but they will still be drawn to some vices and will struggle with making the right choice. An incontinent agent will also struggle with their attraction to vice, but they lack sufficient strength of will. In losing their struggle between virtue and vice, they may feel shame and inadequacy. The vicious, on the other hand, have adopted a false ideal of what is good, and so they will pursue vice with abandon, mistaking it for the good life. Social media and gaming companies are designed to pander to the vicious and pull in all but the most virtuous users through clever tactics of manipulation overcoming the users' strength of will. Cultivating the virtues and strengthening our will can be an effective means of overcoming the mass of mental malware with which we are daily confronted.

Marshall looks at mental malware through a deontological lens. Any attempt at influencing others must, first, be directed towards an ethical purpose. In deontological ethics, it is forbidden to use others as a *means to an end*. We cannot trample over others to fulfil our own interests, which is what many social media companies and marketers are doing. Second, any type of ethical influence must be based on honesty and rational discourse. Mental malware seeks to appeal to what Kahneman calls 'System 1' thinking – the emotive and automatic responses (and therefore quick and easy) that we initially make to new information. However, any ethical attempt at influence must also engage our 'System 2' thinking – our conscious, rational, and deliberative (and therefore slow and difficult) ways of thinking. Those who would deploy algorithms ethically must be honest and transparent about how the algorithms are operating. Finally, we need to prioritize the rational deliberative process of making decisions based on good information and empirically sound data, over the quick, emotional – and therefore easily manipulated – responses that are currently driving much engagement on social media.

Case 7 – AI and Nonhuman Animals

Humans are not the only living beings whose interests are affected – for good and for ill – by ML systems. In this chapter, Singer and Tse present their research on the ways that AI algorithms impact animal welfare. First, they discuss the various effects that search-engine results and recommendation algorithms can affect how we think about – and consequently how we *treat* – animals. Algorithmic bias in search engine results and content recommendations can present us with content and advertisements that influence how much animal products we consume while obscuring real-world animal cruelty and desensitizing users to these harms. Language models can reinforce speciesist meanings of language that objectify and denigrate nonhuman animals. This has a large effect on animal welfare.

Second, they discuss the use of AI in factory farms. ML models are deployed in the factory farm industry to gather information on farmed animals to maximise profits. How much do disease and mortality maximize profit? How much should animals be fed to balance growth with keeping costs low? They also raise the important issue of how we identify and interpret nonhuman animals' behaviour and subjective states of well-being when we are looking through the lens of our own, human, perspective. What does it mean for AI to step out of our human mindset, and adopt a set of nonhuman values and perspectives? The future welfare of nonhuman animals depends on how we resolve these ethical issues.

Sinclair provides a Jewish ethical perspective on the duty to treat animals in an ethical way. While humans are superior to all other beings, the first people were vegetarian, and later when they became corrupted, they were allowed to eat meat. The notion of preventing cruelty to animals is deeply embedded in Jewish ethics including allowing the working animals to rest on Shabbat and enjoy their free time.

Marshall discusses the moral status of animals in deontological ethics. All versions of deontology recognize the moral significance of animals' rights, although they differ on whether animals are morally important in the same way that humans are. If so, then using animals as mere means to our ends would be morally impermissible. He also makes the important point that our failure to adequately respect the moral claims of animals can lead us to fail to respect the moral claims of others more generally. Indifference to the suffering of others – any sentient others – must be avoided by moral agents.

Murungi provides a commentary on the moral rights of animals from the perspective of African ethics. He explores the role played by decolonization in the ethics of AI, and whether we can develop an ethics of AI that seeks to realize the communal nature – the 'we' – that forms the heart of Ubuntu ethics. Students and architects of these technologies often fail to "see themselves as children of AI or as mothers and fathers of AI." Instead, they should be encouraged to think about what it means to be a moral agent, and what constitutes well-being. This opens up a space for a liberatory rather than an oppressive AI – one that also promotes the well-being of animals for "we cannot recognize animals as bearers of moral rights if we inadequately have a sense of who or what we are."

All the commentators in this book provide a way forward for data scientists to engage ethically when designing and using data and AI systems. Indeed, engaging with the comments of the contributors would certainly foster the kind of wisdom for which Kelleher has advocated, and that will do so much to help in navigating the world of algorithms and AI. The importance of data in this age cannot be overstated. This places great power in the hands of data scientists and as the ancient adage states, "With great power comes great responsibility."[21] We truly hope that this book provides some valuable tools that will assist us all in handling this great responsibility with reason, compassion, and wisdom.

Notes

1. Merriam Webster, "ethic," *Merriam Webster Dictionary*, 2022, accessed July 1, 2022, https://www.merriam-webster.com/dictionary/ethic.
2. Many thinkers, for example, are beginning to look at data science and machine learning through the lens of decolonialism as an ethical and political standpoint. See *e.g.*: Nick Couldry and Ulises A. Mejias, *The Costs of Connection: How Data is Colonizing Human Life and Appropriating It for Capitalism* (Stanford: Stanford University Press, 2019); Michael Katell, Bernease Herman, Corinne Bintz, Meg Young, Vivian Guetler, Daniella Raz, Dharma Dailey, Aaron Tam, and P.M. Krafft, "Toward Situated Interventions for Algorithmic Equity: Lessons from the Field," in *Conference on Fairness, Accountability, and Transparency (FAT* '20)*, January 27–30, 2020, Barcelona, Spain, ACM, New York, doi: https://doi.org/10.1145/3351095.3372874; Shakir Mohammed, Png, Marie-Therese, and William Isaac, "Decolonial AI: Decolonial Theory as Sociotechnical Foresight in Artificial Intelligence," *Philosophy & Technology*, 33 (2020): 659–684, doi: https://doi.org/10.1007/s13347-020-00405-8.
3. Margaret A. Boden, *Artificial Intelligence: A Very Short Introduction* (Oxford: Oxford University Press, 2018), 1. General Artificial Intelligence refers to machines that possess a general cognitive capacity for intelligence, much as humans do; such machines do not currently exist, and the theoretical possibility of such machines remains controversial.
4. Boden, *Artificial Intelligence*, 39–40.
5. John D. Kelleher and Brendan Tierney, *Data Science* (Boston: Massachusetts Institute of Technology, 2018), Preface. Loc. 2/34.
6. Luke Dormehl, *Thinking Machines: The Quest for Artificial Intelligence and Where It's Taking Us Next* (New York: Tarcher Perigree, 2017), 17.
7. Dormehl, *Thinking Machines*, 17.
8. Kelleher, *Data Science*, Preface. Loc. 4/34.
9. Dormehl, *Thinking Machines*, 32–33.
10. Dormehl, *Thinking Machines*, 20.
11. Kelleher, *Data Science*, Chapter 1. Loc. 168/346.
12. Dormehl, *Thinking Machines*, 50–51.
13. Kelleher, *Data Science*, Chapter 4. Loc. 20/488.
14. Kelleher, *Data Science*, Chapter 4. Loc. 35/488.
15. Kelleher, *Data Science*, Chapter 1. Loc. 328/346.
16. Kelleher, *Data Science*, Chapter 2. Loc. 143/245.
17. Kelleher, *Data Science*, Chapter 1. Loc. 185/346.
18. See: Julia Angwin, Jeff Larson, Surya Mattu, and Lauren Kirchner, "Machine Bias," *ProPublica*, May 23, 2016, accessed July 1, 2022, https://www.propublica.org/article/machine-bias-risk-assessments-in-criminal-sentencing.

19. Synthetic media – also called 'deep fakes' – use the power of ML systems to create synthetic media, including audio, photo, video, and written text, that appear authentic. See: Tracey Dowdeswell and Nachshon Goltz, "The Clash of Empires: Regulating Technological Threats to Civil Society," *Information & Communications Technology Law,* 29, no. 2 (2020): 194–217.
20. Retraction Watch, "The Top Retractions of 2021," *The Scientist*, December 21, 2021, https://www.the-scientist.com/news-opinion/the-top-retractions-of-2021-69533.
21. Anna Wichmann, "The Sword of Damocles: Why Great Power Comes with Great Danger," *Greek Reporter,* May 25, 2022, available at: https://greekreporter. com/2022/05/25/sword-damocles-power-ancient-greece/. This ancient adage dates back to the 'Sword of Damocles', as told by Cicero, in which the tyrant Dionysus places an obsequious courtier, Damocles, in the position of ruler for a day. In addition to letting Damocles taste the luxuries and pleasures of riches and power, Dionysus dangled a sword over Damocles' head – suspended only by the slenderest of threads. On the sword was inscribed this famous adage.

2 Introduction to Ethical Approaches in Data Science

> Knowledge of physical science will not console me for ignorance of morality in time of affliction, but knowledge of morality will always console me for ignorance of physical science.
>
> ~ (Blaise Pascal 1624–1663)

Introduction

Machine learning technologies are encroaching into the lives of ordinary people around the world. Users of these technologies – some willingly, some unwillingly – have ethical lives that are informed by approaches not represented among Western philosophies. This is the lens through which these users of technologies come to understand them.[1] We have therefore included several non-Western approaches to ethics. These are worth knowing for designers, both so that they can deepen their own exploration of ethics, and so that they can better understand how their technologies are going to be interpreted, adopted, used, and (inevitably) regulated.

We have been fortunate to receive commentaries from leading scholars in the fields of deontological ethics, consequentialist ethics, and virtue ethics, as well as from Ubuntu ethics, Buddhist ethics, Jewish ethics, and Indigenous ethics. We hope that this will give the reader a broader base to begin to think about machine learning technologies from different perspectives and to understand how they will be adopted by and function in societies around the world. Each of these ethical approaches is briefly summarized below.

Consequentialism and Utilitarianism

By Peter Singer and Yip Fai Tse

Consequentialism is a family of theories that hold the view that whether an action is right or wrong depends on its consequences, or in other words, the states of affairs the actions cause. Utilitarianism, in its classical form, is the consequentialist theory that focuses exclusively on pain and pleasure, or happiness and misery, as the only morally relevant consequences for determining how the consequences of actions are to be evaluated.

DOI: 10.1201/9781003293125-2

Here, it is important to emphasize that utilitarianism is not only about assessing the rightness or wrongness of actions, but also about assessing the goodness and badness of states of affairs, considered impartially. Specifically, Utilitarians hold that all sentient beings – those that can experience pain and pleasure – must be considered, and their similar interests[2] should be given similar weight. Put together, Utilitarianism is therefore the view that an action is not only right but also morally required when it brings about the greatest possible net surplus of happiness over misery (or of pleasure over pain); and also that an action is wrong and impermissible when it does otherwise.

Common Objections to Utilitarianism

A common objection to Utilitarianism is that it directs us to perform clearly immoral acts. In *The Brothers Karamazov*,[3] Dostoevsky has Ivan challenge his brother Alyosha to imagine whether he would accept torturing a baby to death in order to bring happiness to all of humanity. Ivan's challenge has become a famous objection to Utilitarianism. Setting out the structure of Dostoevsky's objection formally better illustrates the issue:

Premise 1. If Utilitarianism were true, it would tell us, correctly, which acts are right and which are wrong.

Premise 2. Utilitarianism tells us that if torturing an innocent child to death would bring about better consequences than any other action, then torturing an innocent child to death would be the right thing to do.

Premise 3. Torturing an innocent child to death is always wrong.

Conclusion: Utilitarianism is false.

Many objections to Utilitarianism are similarly constructed: a surgeon contemplates whether to secretly ensure that an operation fails so that the patient dies and her organs can be used to save the lives of four patients waiting for donations of essential organs. Such examples do not reflect our knowledge of how the world works. Ivan did not explain how torturing the child would bring lasting happiness to everyone else. The organ transplant example fails to consider that if what the surgeon has done becomes known, this might lead to consequences that far outweigh the intended benefits – people might become distrustful of doctors and reluctant to seek medical help. How can the surgeon be entirely sure that she would not be caught? The assumption that torturing an innocent child is always wrong relies on an intuition that evolved to respond to circumstances we are likely to encounter. When we are dealing with fantastic examples, Premise 3 is therefore doubtful and cannot be relied upon as a basis for refuting utilitarianism.

Another major objection is that pain and pleasure, or welfare states, are hard to measure. Utilitarians have three main responses to this objection. Firstly, this is not a problem limited to Utilitarianism. Any moral theory that gives some weight to welfare will also suffer from the difficulty of measuring the welfare of those affected by our actions – and a moral theory that ignored all such welfare considerations would be extremely implausible.

Secondly, even though it is difficult to measure pain and pleasure precisely, the preferences of people, and to some extent animals, can be observed, tested, and ranked to reveal their priorities. In some studies, psychologists test people's willingness to pay to avoid[4] or endure[5] certain levels of pain. These revealed preferences, although not what classical Utilitarians regard as the ultimate good, are nevertheless useful proxies that give us some idea about the underlying pain and pleasure.[6] Another model using revealed preferences, the quality-adjusted life-year, or QALY, is framed around the idea that a year of life with impaired functionality or health is not as good as a year in normal health. To find out how much less good it is, researchers ask people to imagine themselves with various impairments to their health (sometimes pain itself), and then ask them how many years of life they would give up in order to have that impairment cured. This method is now used globally by health economists, medical researchers, and policy-makers.[7]

Finally, in many cases, the right action is clear even without measurement. For example, a doctor prioritizing the order of patients can clearly see that a heavily burned patient is in much more pain and at much higher risk of dying than someone suffering from a cold, and should therefore prioritize the burn patient. Or, if someone asks you how to get to the nearest vegan restaurant and you know the answer, you will very likely maximize utility by providing her with the information she is seeking, rather than not responding at all, or lying to her. In both cases no "measurement" is needed.

While it is true that there are cases where things are not nearly as clear, after carefully analyzing evidence, reasonable choices can nevertheless be made. An important point to consider here is that not only can a significant portion of *decisions* under Utilitarianism be made without measuring pleasure and pain, but what is *at stake* in this world can also usually be determined without direct measurements of pain and pleasure. Global poverty (which causes hunger, thirst, diseases, and "poverty traps"), factory farming, and pandemics are pertinent examples of issues that undoubtedly cause enormous amounts of suffering to an enormous number of individuals.

Advice for Properly Applying Utilitarian Principles

Think Broader and Longer

We agree with John Stuart Mill, an early utilitarian, that we should "regard utility as the ultimate appeal on all ethical questions; but it must be utility in the largest sense". By "largest", we mean that all relevant consequences should be considered, regardless of their timing, physical distance, kinship, and other morally irrelevant characteristics such as gender, race, and species membership.

Arguably, time is one of, if not the, most controversial of these to be declared morally irrelevant. Time discounting is often taught and applied in the fields of economics and machine learning, and their notions about time preferences are different from those of Utilitarianism. In economics, time discounting is used to capture people's general preferences for a nearer enjoyment than a farther one. In machine learning, particularly in reinforcement learning, the "discount factor" (or γ) is the variable that determines how much a reinforcement learning agent discounts rewards depending on how far in the future they are. It is used for various reasons: it prevents the reward from becoming infinite, it captures the intuition that there is generally more uncertainty in the future, and it sometimes improves the stability of the agent.[8] These practices of using discounts might give one the impression that time preference is justified. But we need to distinguish between pure time preference due solely to time itself and discounting time for other reasons. Utilitarians, because of their impartiality, do not hold pure time preferences but accept that other practical reasons, such as uncertainty, are justified reasons for discounting. For example, torture in 100 years is equally bad as torture that brings the same amount of pain now, but if it is less certain, we may discount it for that reason.

Let us try to apply these principles to artificial intelligence (AI) and data science. In deciding to launch a product, for example, one should not only consider the impact it might have on its users, but also how the broader society of individuals (in some cases, even nonhuman animals) could be affected, both in the short-term (such as within the product's lifecycle) and in the long-term. Questions such as these should be asked: Will this product capture, and thereby reinforce biases, culture, ideologies, virtues, or other values in society? Will this product eliminate a hugely valuable industry, or will it delay or prevent the elimination of a morally objectionable industry?

Use Expected Values to Make Decisions

Using expected value theory in decision-making is foundational in decision theory, economics, and data science. But it should also be applied to ethical theories, especially maximizing ethical theories such as Utilitarianism. The surgeon example in the previous section shows why low-probability-high-stakes scenarios are important. No matter how carefully the surgeon tried to keep her act secret, she could not reasonably conclude that the chance of the secret being exposed is zero. Considering the potentially catastrophic effects of being discovered, the surgeon ought to conclude that it would be wrong to perform such an act.

While calculating expected value is often straightforward, it is often improperly applied or even not applied at all, due to human cognitive biases. Probability neglect is a cognitive bias people show toward uncertainties, especially that of small probabilities, which they have the tendency to either neglect entirely or exaggerate greatly.[9] A study confirmed this by finding that people placed either a high or a very low value on mitigating risks of "rare, high-impact events".[10] We need to look no further for evidence of a collective neglect of "rare, high-impact events" than the fact that without legal compulsion, many drivers and passengers in cars would not do up their seat belts. This pitfall in human thinking is worrying, as events with low probability and the high impact often have large expected values, either negative or positive.

Another bias that might further affect people's ability to estimate the expected values is scope neglect. Studies found that people do not adjust their valuation of an issue in proportion to its scale.[11] For example, one study asked three groups of subjects about their willingness to pay to save 2,000, or 20,000, or 200,000 birds from drowning in uncovered oil ponds. The respective means were $80, $78, and $88, and the median answers were all $25.[12] If people's valuations of certain outcomes do not scale properly, the expected values won't either.

In Choosing Charitable Projects, Choose Effective Ones

As people commonly give to charity on the basis of impulse and emotion, rather than research into the effectiveness of the charity, they often support ineffective charities and causes.[13] In response, Effective Altruism, a recent global movement, emphasizes the importance of effective altruistic behavior, either in the form of monetary donations or time. One of the movement's most important contributions has been to assess the available evidence in order to discover and publicize the causes and organizations that are most highly effective in doing good.

Applying the same principle to the fields of AI and data science, companies and scientists searching for ways to do good should prioritize the approaches that show evidence of being highly effective or having high expected value. Examples of this research, and sample analyses on the effectiveness of various causes, careers, and charitable programs can be found on the websites of such organizations as Giving What We Can, 80,000 Hours, The Life You Can Save, and Animal Charity Evaluators.[14]

Deontological Ethics

By Colin Marshall

Deontological approaches to ethics focus on a cluster of related ideas: respect, autonomy, rights, and refusing to treat humans (and perhaps other beings) as though they were mere things or means to other ends.

A classic illustration of the deontological approach involves the following scenario: a doctor has five patients each in urgent need of organ transplants. A trusting and healthy person walks into the doctor's office, and the doctor could kill the healthy person and harvest their organs in order to save the five patients. Even if the doctor could do this without being detected, many people judge they should not. This judgment is easily expressed in deontological terms: as a failure of respect on the doctor's part, as a violation of the healthy person's rights, or as the doctor using the healthy person as though they were a mere a thing (an organ container).

The deontological approach is often contrasted with consequentialist approaches, which recommend whichever action yields the best outcome. And, if the details were filled out the right way, a consequentialist approach to the doctor case could recommend that the doctor sacrifice the one healthy person to save the five patients. In practice, however, the verdicts of deontological and consequentialist ethical approaches frequently coincide. After all, in any *realistic* version of the doctor case, there could be no guarantee that the murder would remain secret. Given that, consequentialism would recommend factoring in the huge risks involved, such as decreased trust in medical professionals (resulting in sick people failing to seek necessary help) and the likely devastating psychological impact on the doctor (whose guilt and trauma might detract from their future work). As a result of such considerations, many consequentialists have held that the best consequences would be secured if people generally approached decision-making from a deontological perspective. For that reason, we can expect many deontological evaluations of ethical cases to agree with consequentialist (and other) evaluations, even if the different approaches emphasize different factors.

In the commentary below, the central deontological notion will be that of *respect*, together with two notions that emerge from respect: *impartiality* and the *refusal to others as mere means*. We can briefly consider each of these here.

There are many varieties of respect, but the relevant form of moral respect is giving serious consideration to others' needs and projects. Such moral respect can and should often impact action: if we seriously consider someone's needs, we would typically not act so as to undermine those needs. Yet there can be failures of respect even when no action results, as in disrespectfully laughing at others' failures.

Respect is hardly our default attitude toward others. Instead, we are frequently drawn toward disrespectful partiality, focusing on our own needs and projects (or those of some group to which we belong) at the expense of others' needs and projects. For that reason, respect calls for impartiality in our dealings with others. It is difficult to tell when we are being appropriately impartial since we often construct impartial-*sounding* rationalizations for selfish behavior (consider the dangerously flexible cliché, "our product will make the world a better place"). For that reason, deontology often recommends certain forms of reflection to help detect partiality – see the commentary on Case 2 – "Machine Models in Court" for one example.

While impartial respect might be an absolute standard, partiality comes in degrees. Someone can be partial toward their social group, for example, while still giving *some* consideration toward other social groups. At an extreme, however, partiality involves giving *no* serious consideration toward certain other beings' needs and projects, and so treating them as nothing more than tools for achieving certain goals. If a social media platform, with the sole aim of maximizes profit, entices its users into forms of harmful engagement, they are treating their users as mere means for achieving profit (see the commentary on Case 6 – "Mental Malware" for more). Similarly, if a factory farm treats animals as mere sources of meat, then it is treating them as mere means (see the commentary on Case 7 – "Animals and AI"). Such an attitude constitutes a complete failure of respect.

Whether or not deontology captures the ultimate ethical truth in some sense, many people care about respect, rights, and impartiality. There are therefore both economic as well as moral reasons for taking the deontological approach seriously.

Virtue Ethics

By John Hacker-Wright

Virtue ethics is an approach to ethics or, more accurately, a family of approaches, that grounds its guidance of action on the states of character that a human being needs to live well. It tells us to develop and exhibit good states of the character called virtues, and to avoid developing and exhibiting bad states of the character called vices. The most prominent strand of virtue ethics in the Western academy today is informed by the ancient Greek philosopher Aristotle (384–322 BCE), but there were and continue to be many versions of virtue ethics. Hence, one can find, for example, Confucian and Buddhist versions of virtue ethics. The view I will expound on what follows and that informs my approach to the cases is Aristotelian virtue ethics.

When thinking of a good person, you might think of someone with qualities such as courage, compassion, honesty, and the like. These are putative virtues. Whatever qualities we think someone needs to live well in a certain domain of human life comprise our conception of the virtues. While there is no definitive list of virtues, there is significant convergence over qualities such as courage, honesty, justice, and wisdom. Virtue ethicists try to derive a criterion of right and wrong in action from the virtues or the virtuous person. One prominent formulation goes: an action is right if and only if it is what a virtuous person, acting in character, would do in that situation.[15] Note that we can follow this even if we are not, ourselves, virtuous, provided we have some level of insight into what the virtuous agent would do and sufficient self-control to carry through with acting as the virtuous person would act. If our desires are too disordered, we may not be able to carry through with virtuous intentions and may even act worse as a result of trying to act as a virtuous agent would act![16] In that case, our moral agency is compromised by the weakness of will, and a virtuous person would advise a remedial course of action. Our goal remains to become able to act as the virtuous agent would act.

Perhaps the most significant difference between virtue ethics and other approaches to ethics is that virtue ethics embraces non-codifiability: there are no exceptionless rules that could capture what it is to act well across the variety of situations we face in human life. There may be rules that capture loose generalities, patterns of action, and reasoning characteristics of virtuous people, but these cannot be applied unthinkingly. In other words, some level of moral understanding is necessary to apply them. This might seem like a weakness of the theory, but on the other hand, competing theories commit themselves to deeply counterintuitive and sometimes morally disturbing views of right action based on exceptionless rules: for example, Immanuel Kant's deontological view is infamously committed to an exceptionless position on never telling a lie, even if it would save lives to do so. Virtue ethicists, by contrast, may hold that the human need for trusting relationships makes honesty a virtue and yet contend that we can uphold our commitment to honesty while allowing circumstances that are calling for deception. For example, it may be appropriate to lie when we are pressed for information about someone's whereabouts by someone with obvious murderous intentions. The absence of exceptionless rules may also be an asset for virtue ethics when it comes to dealing with emergent technologies, which change the nature of our practical lives.[17]

Since the virtues are at the center of virtue ethics, it is crucial to understand what they are. Some virtues are excellences of our desires and feelings, whereas others, such as practical wisdom, are primarily intellectual excellences. Courage, for instance, concerns our desire for safety and it is exhibited when our feelings of fear and confidence are such that we feel fear only in the face of what is genuinely dangerous and to the extent that it is dangerous. Aristotle explained the idea of virtue by appealing to his famous "doctrine of the mean."[18] In a courageous human being, the feelings of fear and confidence are in a medial state between an excess and a deficiency. Someone who feels excessive fear will flee from danger and fail to achieve something worthwhile. We label such people cowards as they exhibit the vice of cowardice.

Someone who feels too little fear may act recklessly and face injury or death in vain efforts that should have been avoided. It is characteristic of the Aristotelian approach that fear, among other feelings, is something that is necessary to live well; after all, when I feel fear, I register the value of my own life and physical integrity in a way that motivates action. Still, I may overvalue my life and physical integrity.

In Aristotle's view, there are more important things than my own life and physical integrity, such as the liberty of my city-state and the safety of my friends and family. Hence, on his view, it is a good thing to risk death if there is a non-trivial chance of achieving some such end. Another aspect of Aristotle's view is that one cannot exhibit courage unless one is facing fear for the sake of a worthwhile end. A thief who faces danger for the sake of stealing is not courageous. Although their character is such that they are not prone to feeling excessive fear, still, this state of character is not an excellence in them.

Their viciousness in another domain – say, acquisitiveness – undermines their ability to act courageously. This is another aspect of Aristotle's doctrine: he advocates an idea called "the unity of the virtues" which states in its strongest form that to have one virtue we must have them all. In a somewhat weaker and more plausible form, the idea is that any vice undermines the ability to exhibit any virtue. Assuming that there are states intermediary between virtue and vice, this leaves room for being less than fully virtuous in some areas without necessarily undermining our virtue in other areas.

Staying with the virtue of courage as an example, we can wonder whether it is good to be courageous. After all, if it requires that I risk my life for the sake of my city-state, perhaps it is better for me to be cowardly. But note that this accepts the cowardly perspective on the world: that it is better at any cost to survive. The courageous person sees the world differently: survival is not good when it comes at the cost of the freedom of one's city-state or the death or enslavement of one's friends and family. Are we left, then, with the juxtaposition of these two different perspectives, or is there some objectivity to the courageous person's perspective? I believe that the courageous person's perspective is superior because courage is a trait that human beings need to live well in a world of danger. We human beings need to be able to carry through with goals even when faced with dangers. This view is a version of ethical naturalism that many Aristotelians embrace: that goodness in humans is a function of the sort of animal that they are.[19] The virtues perfect human rational and appetitive powers, and it is an objective matter, of which traits do that.

Aristotle wrote in an age with a very different social structure as well as with different technologies. One might wonder whether his views are relevant to our times. Certainly, no Aristotelian virtue ethicist writing today accepts his views without modification. There is an over-emphasis on martial virtue of courage and a heavy aristocratic bias to his conception of the virtues; his political views are tainted with the misogyny and racism of his time. But his philosophical framework nevertheless still yields insight. In addressing questions of technology and data science, Aristotelian virtue ethics emphasizes an examination of the impact of the virtue on our character: how does the use of a new technology impact our desires and thinking? If a technology is making us think or feel something as characteristic of a vicious agent, then this is grounds for ethical criticism of the technology. Hence, the focus is on how well we are living with the technology. We do not have to find a dramatic impact on society or the violation of duties to raise moral qualms about a technology; we can approach ethical criticism of technology by taking stock of distortions and its effects on our thoughts and feelings. New technologies may make new ethical demands on us. In such cases, it is a question whether a new virtue is required, or whether it is simply a matter of thinking about a traditional virtue within a new context. My view is that new labels tend to repackage aspects of traditional virtues, and that there is no harm and may be some benefit in doing that, since it may help us to think more carefully about the situations we face. In short, Aristotelian virtue ethics provides a flexible framework for thinking about how well we are living with new technologies, and it need not be tightly tied to Aristotle's ancient views about the virtues.

One final aspect of Aristotelian virtue ethics that is important to bear in mind is the idea, found in Aristotle, that knowledge of virtue and the conditions of its development to be important to politics. Virtue ethics is misunderstood if it is assumed that we as individuals are on our own to develop and exhibit traits that we need to live well under whatever circumstances. Instead, virtue ethics is or ought to be concerned with weighing in on the social conditions that are necessary for human beings to live well. This is especially important in considering the impact of new technologies. They may undermine (or, to be more optimistic, enhance) our ability to bring our desires into line with our conscious goals, and thereby undermine our efforts to develop the virtues. On the Aristotelian view, the development of virtues requires a process of habituation, that is, a process of acting as the virtuous agent would act, perhaps against our inclinations, until we come to enjoy acting in that way and can do it reliably on our own. This process happens in a community and begins before we are capable of conscious, intentional action. Hence, it matters very much what sort of community we start from, and much of what virtue ethics will have to say about technologies will come from examining the impact of technology on communities from a standpoint of whether it enables us to develop self-control and excellences of character.

A Provisional Perspective on African Ethics

By John Murungi

In what follows, one should expect a provisional perspective on African ethics. Provisionality matters as it leaves room for other perspectives. Moreover, it alerts the reader to the fact that what is said about African ethics does not exhaust what can be said about it; there is more to be said. African ethics lacks the solidity of a granite rock. Even such a rock, regardless of how hard it is to penetrate, remains penetrable, since nothing in nature is essentially impenetrable. What is true of African ethics is equally true of any other branch of ethics.

African ethics is not a datum placed on a calabash for warehousing or for transmission from one person to the next. It is not placed there for contemplation or for reflection by any one person. It is in the process of being what it is. One should not exclusively rely on an empirical perspective unless such a perspective is understood as provisional. The prejudice of empiricism solidifies data so that it can be subject to an empirical method of verification. This prejudice robs data of its essential processional and provisional aspect.

Aimé Cesáire, a Caribbean of African descent from Martinique, was a member of the French Communist Party. In protest against excluding racism as a fundamental category in the struggle against exploitation and oppression, he wrote a resignation letter in which he said, "... there are two ways to lose oneself: walled segregation in the particular or dilution in the 'universal'."[20] This is an important insight if one is to do justice to the understanding of African ethics. African ethics is a singular ethics as is the case with any other branch of ethics. It should not be mistaken for or confused with any other branch of ethics. It has indestructible autonomy. It must be understood in its own terms. It ought not to be diluted in the "universal." Inescapably and at the same time, it is not walled in segregation in the particular. It is universal, not in a false sense of universality but in true universality. True universality does not dispense with true particularity. It dispenses with false particularity. It is instantiated in the true particularity. True particularity is itself instantiated in true universality. The distinction between true universality and true particularity is an ongoing fundamental problem. The problem is intrinsic in what is honestly said about African ethics.

In speaking about African ethics, one must address what is African about ethics. What is African about African ethics is not exclusively a matter of physical geography or a matter of biology. If this was not the case, African ethics would be a matter of physical geographers or a matter of biologists. When ethics is at stake, one must pay attention to geo-ethicists or to bioethicists bearing in mind that ethics is not a physical science or a bio-physical science. Ethics is elemental and does not depend on what is extra-ethics for its being. What is African about African ethics has an irreducible aspect of true particularity and, at the same time, it has an inescapable aspect of true universality.

I propose that the particular-universal aspect of African ethics lies in the African concept of Ubuntu. This is claimed not at the expense of other concepts of African ethics – both existing and the ones to come. Ethics, whether African or non-African, whether particular or universal, is about well-being. In Indigenous African societies, societal well-being is communal well-being. It is well-being that has a place for individual well-being as well as group well-being. In either case, such well-being is not at the expense of society's well-being. This broader sense of societal well-being rests on the African concept of Ubuntu or on other similar concepts such as Ujamaa, or on what used to be referred to as African Socialism. In the Ubuntu mode of human being, it is claimed that "We are, therefore I am." In Ubuntu, it is recognized that, in the pursuit of well-being, there is a place for an individual or a group to pursue well-being but not at the expense of the pursuit of the well-being of society. It is also recognized that, in pursuit of well-being, an individual or a group could act in a manner that is contrary to societal well-being. In such cases, there are societal controls to minimize the threat to society's well-being. The legitimate pursuit of society's well-being, however, cannot be undertaken at the expense of legitimate individual or legitimate group pursuits. Legitimacy in either pursuit is a matter of ongoing deliberation. Moreover, borders of the individual, group, or society are permeable. They are subject to change, and they too call for ongoing deliberation.

It is important to remind ourselves that African ethics is not subject to ethnography, or ethnology. It is not ethnic or tribal ethics. Today, in the light of colonization and racism, it is of utmost importance deracialize and to decolonize the concept of African ethics. If there is going to be an honest and a true version of African ethics, decolonization of the language of this ethics, both within and outside Africa, is necessary. It must not be forgotten that African ethics is what it is: African ethics. African ethics ought to be defined ethically. It is a human ethics – a human ethics defined ethically. It must be borne in mind that the process of definition is essentially incomplete, and, in the foreseeable future, it remains as such. Moreover, it should not be restricted to the domain of the human. The domain of the human extends to the domain of the non-human. It is what it is by being encompassed by the domain of all that is. Its frontiers are ever expanding. It is the acceptance of membership of Africans in this broader domain that lead European explorers in modern history to the radically erroneous belief that Africans were savages or uncivilized. Correcting this error is an essential aspect of African ethics.

Because ethics is implicated in the well-being of society, it seems proper that sociology is indispensable in the study of ethics. As sociology is the study of society, the study of ethics is embedded in sociology. Moreover, since society is politically secured and its interests are promulgated and pursued by the state (the political), ethics is fundamentally political. Differently stated, ethics is subject to political sociology. In conventional Euro-Western ethics, the multilayered architecture of ethics is rarely recognized. In the Indigenous African context, this architecture is recognized. The sense of the political in Africa is radically different from the modern European concept of the political. In the African context, the politico-ethical is akin to the Ancient Greek political sociology. In his book on politics,[21] Aristotle tells us that by nature man is a political animal. Modern Euro-Western sense of the political deviates from this conception of the political. It seems that Indigenous Africans share the Greek sense of the political and, indeed, there are other societies that share it. Euro-Western modernity appears to be in flight from this political ethical sense. It is subversive of the ethical of the Indigenous Africa. To this extent, the Euro-Western sense of the ethical is in conflict with Indigenous African sense of the ethical.

Buddhist Ethics

By Peter Hershock

Ethics covers a considerable amount of conceptual ground. It has been equated with moral reflection on how 'we' do or do not do things from within a historically, culturally, and geographically located community. But it has also been equated with rationally justified derivations of principles of conduct deemed applicable across and among communities, not just within them. Thus, ethics can involve everything from explications of what is involved ideally in being a "good" person, to that of the practical meaning of "acceptable" agency within a profession or as citizens of a nation or the world.

I approach ethics operationally, defining it minimally as the *evaluative art of human course correction*: the art of intelligently enacting the results of discriminating collaboratively and qualitatively among our values, aims, and interests, and our means for realizing them. For me, it is an art crucially informed by contemporary elaborations of Buddhist concepts and practices.

Buddhism emerged in the Himalayan foothills of South Asia some 2,600 years ago, at roughly the same time as the philosophical and political traditions of the Mediterranean and Sinitic worlds. Those traditions grappled with such fundamental questions as: what is real? what is good? what is the place of humanity in the cosmos? and, how should society be governed? Buddhism arose in therapeutic – rather than theoretic – response to two different, but equally fundamental questions. What are the causes and conditions for experiencing *duḥkha* or suffering, conflict, and trouble? And, by what means can we dissolve these causes and conditions?

Buddhist responses to these questions rest on two key insights. First, all things arise and persist interdependently. Strongly stated, relationality is more basic than things related. All things are a function of relational differentiation, and each thing ultimately *is* what it *means* to and for all others. Second, our cosmos is self-organizing and karmically-structured. It is a cosmos in which consistent patterns of values, intentions, and actions result in consonant patterns of experiential outcomes and opportunities.

The Buddhist art of human course correction has the aim of realizing freedom from *duḥkha*-engendering relational entanglements, most proximally through resolving conflicts among our values, intentions, and actions. This depends on critically assessing habits of thought, speech, and conduct, and on realizing the freedom of attention and freedom of intention required to revise, resist, or dissolve those habits as needed to be no longer constrained by karmic entanglements and compulsive presence. Significantly, the aim of Buddhist practice – realizing nirvana – is not prescribed or positively defined. Instead, it has traditionally been glossed metaphorically as cooling down or extinguishing the fires of craving, aversion, and ignorance. This has important ramifications for Buddhist ethics.

Among the simplest formulations of Buddhist practice is that, for the purpose of ending conflict, trouble, and suffering, one should see all things as impermanent and without fixed essence or identity, and as liable to triggering relational blockages and entanglements if seen otherwise. This practice opens possibilities for being present and acting in ways that are *kuśala* and eschewing those that are *akuśala*. Although it is often translated as "wholesome" or "skillful," *kuśala* carries the normative force of a superlative. It directs us to envision, not just avoiding bad and mediocre conduct, but also merely good conduct, and orienting ourselves instead toward realizing virtuosic conduct and relational outcomes. In short, Buddhist ethics is thus not goal or destination-oriented. It is an open-ended and improvisational art.

Buddhist ethics can, with some justification, be seen as involving elements of the virtue-based, duties-based (Deontological), and consequence-based (Utilitarian) approaches to ethics that have become dominant in Western philosophy, as well as those central to care-based (Feminist). Yet, Buddhist relational ontology distinctively orients evaluative attention away from independently existing ethical agents, patients, and actions, and toward relational quality. Moreover, while Buddhism's emphasis on relational virtuosity commits Buddhist ethics to being context-specific, it differs from Western situational ethics which evaluates actions in terms of their proximate or near-term outcomes. What matters ethically are not simply the immediate consequences of an action, but also the medium – and long-term relational ramifications of intentionally enacting specific sets of values, and their shaping of *volitional opportunities* as well as *experiential outcomes*.

In sum, Buddhist relational ontology radically expands the spatial and temporal scope of ethical evaluation. Rather than focusing on the character of both spatially and temporally "local" agents, actions, and patients, Buddhist ethics directs evaluative attention – at the very least – to show our value – and intention-enacting presences qualitatively affect the networks of internal relations that constitute us as persons in the planetary community with all other sentient beings and their sensed environments.

This expansion of evaluative scope arguably entails a metaethical commitment to fostering ethical diversity. Especially in addressing the challenges and opportunities associated with the emergence of intelligent technology; what we need is not a new "species" of ethics, but rather an ethical "ecology," the vitality of which depends on ethical diversity or the degree to which ethical differences become the basis of a mutual contribution to shared, relational flourishing. Such an "ecological turn" in ethics is, I believe, crucial for resolving the ethical predicaments emerging as human values and intentions are innovatively scaled up by syntheses of human and machine/algorithmic intelligence.

Jewish Ethics

By Samuel J. Levine

A Jewish ethical approach, like virtually all areas of Jewish thought, is premised upon the foundational sources of Judaism, most prominently the text of the Torah, as interpreted and applied over the course of thousands of years and across geographical locations and societal settings around the world. The text of the Torah, although replete with legal precepts, includes substantial narratives as well, providing context and further meaning to the extensive legal material, while offering ethical lessons of its own. As such, Jewish ethical responses to specific questions may involve a complex consideration of the interaction of various modes of legal and philosophical thought, as developed through millennia of teachings and traditions.

Ethical questions revolving around AI and machine learning may therefore be assessed in reference to a variety of approaches. For example, the status of AI may implicate fundamental principles underlying the nature and definition of human beings. Thus, a starting point for the analysis may very well prove to be the starting point of the Torah, the narrative portions of Genesis that depict the creation of humans as distinct from other living beings. Within Jewish tradition, some philosophers have suggested that this distinction is grounded in the human power of speech, reflecting humans' unique nature as a physical being also created in the image of God. At the same time, humans are thought to be distinct from other divine beings – the angels – in that humans have the power of free will, while angels are compelled to act in accordance with their roles as decreed by God. Whatever the explanation, by all accounts, Jewish ethics, and Jewish law value human life as paramount, thereby outweighing nearly all other concerns. The differing, yet complementary, views of the fundamental qualities of human nature may play a role in evaluating ethical approaches to AI.

Of course, the creation narrative is, by definition, the beginning of the analysis. With the passage of time, technological and societal changes impact law and ethics, as the unanticipated – and the once unimagined[22] – emerge as a reality. Perhaps both acknowledging and anticipating inevitable advances in technology, Jewish sources have explored such issues as the legal status of newly-created beings created by humans and possessing quasi-human qualities, as well as the legal implications of both eternal life and resurrection from death. The emergence of AI may bring some of these issues, which have heretofore remained largely in the realm of theory, to the forefront of practical legal and ethical consideration, serving as yet another among countless examples in which ancient Jewish precepts and principles have been continuously examined, interpreted, and applied to new circumstances.

Indigenous Ethics: Actions as Interactions

By Joseph Len Miller and Andrea Sullivan-Clarke

The question of what an Indigenous moral theory looks like is difficult to answer. First, there is the problem of "Pan-Indianism."[23] Thinking of "Indigenous" peoples as a homogenous group is problematic given the differences that exist between tribes. Second, historically, the philosophical thought of Indigenous people has been seriously underestimated. Most Western thinkers have assumed that Indigenous people were too primitive, or even "savage," to engage in serious reflection regarding abstract topics/questions.[24] This history has lasting impacts on Indigenous philosophy. Not only do Indigenous ideas, even the most fundamental ones, have to be "justified" according to Western standards, but these ideas have to be explained in a context other than the one in which they developed.[25]

That being said, one important common focus of Indigenous ethics is the interconnectedness of all things (e.g., people, the land, non-human animals, past- and future generations, etc.).[26] The universe is a living entity and is understood to be in "constant transition."[27] This provides Indigenous peoples a context that "operates on the principles of balance of harmony."[28] People are born into community and relationships. These include nonhuman relations, such as spirits, rocks, rivers, members of non-human animal species, etc. Each entity with which we stand in relation is different, and so our actions toward our relations will also be different.[29] Instead of providing universal principles for guiding behavior, there are key concepts that provide a foundation and guidance for ethical decision-making. These concepts include those of harmony, reciprocity, gratitude, and humility. Understanding how these concepts relate to one another can help to better understand how to enact these concepts in different contexts. The correct way to live, and act, is then informed by what we know of our relations and our interconnectedness with these concepts.

One of the primary concepts is that of *harmony*. Harmony exists when there is a balance between exchanges and interactions with one's surroundings.[30] Balance and harmony, features of the world that we are born into, offer a guide for ensuring the well-being of our relations and ourselves.[31] Given the interdependence and relationships between all things, every interaction has an effect on the well-being of an individual and their surroundings. In other words, any action is an interaction (i.e., (inter)action). To promote balance in these interactions, an individual has to know how to *reciprocate*.[32] Reciprocation can take many forms (i.e., there is not a singular "right" way to reciprocate), but it should be tailored to the being with whom one's interacting. The aim of reciprocating is to bring balance to one's relationships so that all beings involved can live together peacefully. In order to live together peacefully with one's surroundings, and appropriately reciprocate, one needs to act with *love*, gratitude, and humility.

With these concepts in mind, for any particular (inter)action one should consider the following questions: What action would bring about harmony? How should I reciprocate what I have been given? Am I acting with love, gratitude, and humility? Notice, the answers to these questions are going to depend heavily on one's surroundings and context. Answering these questions in an appropriate way requires having an intimate knowledge of one's surroundings and relationships. For example, knowing how to bring about harmony – i.e., knowing how to reciprocate – in one's relationship with the land requires knowing intimate details about the soil, plant life, bodies of water, weather patterns, interdependence between plants and animals in the area, etc.

Partly as a means of preserving, in the presence of settler colonialism, the ways of life that are structured around these concepts, some political concepts took on a more prominent role in Indigenous ethics. This includes the concepts of sovereignty and reclamation. Since the focus of this collection is ethics, we have prioritized foundational ethical concepts that inform ethical decision-making in Indigenous philosophy. However, given the importance and inspiring recent publicity of interest concerning Indigenous data sovereignty,[33] we would be remiss not to share some resources that highlight how these concepts (sovereignty and reclamation) are used in the *collection* and *use* of data pertaining to Indigenous peoples.

Kukutai and Taylor have recently edited a volume that collects papers in support of "the inherent and inalienable rights and interests of [I]ndigenous peoples relating to the collection, ownership, and application of data about their people, lifeways and territories."[34] Rodriguez-Lonebear and Martinez present an argument in support of "the repositioning of authority over Indigenous data back to Indigenous peoples."[35] Carroll et al. articulate, describe, and present examples of the CARE Principles for Indigenous Data Governance (Collective Benefit, Authority to Control, Responsibility, and Ethics).[36]

In general, Indigenous peoples approach the question of how to live well with humility because they realize that we are only a small part of the universe. We are dependent upon the well-being and generosity of our relatives (i.e., all of our relations) to survive. Dysfunction, chaos, disorder, and the deterioration of the well-being of our relatives creates disharmony and serve as an indication that our actions are misguided and we must change our ways.

Notes

1. Nachshon Goltz, John Zeleznikow, and Tracey Dowdeswell, "From the Tree of Knowledge and the Golem of Prague to Kosher Autonomous Cars: The Ethics of Artificial Intelligence Through Jewish Eyes," *Oxford Journal of Law & Religion,* 9 (2020): 134, doi: https://doi.org/10.1093/ojlr/rwaa015.
2. Although it is commonly thought that utilitarianism is about the "greatest good for the greatest number of people," Bentham and all the major Utilitarians after him have been very explicit in saying that Utilitarians must take into account the pleasures and pains of animals, and indeed of all sentient beings.
3. Fyodor Dostoevsky, *The Brothers Karamzov,* trans. Constance Garnett (Mineola, NY: Dover Publications, 2005).
4. Anderson Chuck, Wiktor Adamowicz, Philip Jacobs, Arto Ohinmaa, Bruce Dick, and Saifudin Rashiq. "The Willingness to Pay for Reducing Pain and Pain-Related Disability," *Value in Health: The Journal of the International Society for Pharmacoeconomics and Outcomes Research,* 12, no. 4 (June 2009): 498–506, doi: https://doi.org/10.1111/j.1524-4733.2008.00457.x.
5. Daniel Read, and George Loewenstein, "Enduring Pain for Money: Decisions Based on the Perception and Memory of Pain," *Journal of Behavioral Decision Making,* 12, no. 1 (1999): 1–17, doi: https://doi.org/10.1002/(SICI)1099-0771(199903)12:1<1::AID-BDM310>3.0.CO;2-V.
6. Revealed preference, or sometimes called inferred preference, is not the same as true preference. Revealed preferences are therefore still proxies for what preference Utilitarians want to measure, albeit much better proxies for preference Utilitarians than they are for classical Utilitarians.

7. See e.g.: China National Health Development Research Center, Report, June 28, 2015, accessed July 2, 2022, http://ncncd.chinacdc.cn/xzzq_1/202010/W02020101084841270 2955.pdf; and Institute for Clinical & Economic Review, "Cost-Effectiveness, the QALY, and the evLYG," *Institute for Clinical & Economic Review* (Website), 2022, accessed July 2, 2022, https://icer.org/our-approach/methods-process/cost-effectiveness-the-qaly-and-the-evlyg/.

8. Vincent François, Raphael Fonteneau, and Damien Ernst, "How to Discount Deep Reinforcement Learning: Towards New Dynamic Strategies," arXiv, December 7, 2015, doi: https://arxiv.org/abs/1512.02011.

9. Amos Tversky, and Daniel Kahneman, "Advances in Prospect Theory: Cumulative Representation of Uncertainty," *Journal of Risk and Uncertainty*, 5, no. 4 (1 October 1992): 297–323, doi: https://doi.org/10.1007/BF00122574.

10. Nicholas Barberis, "The Psychology of Tail Events: Progress and Challenges," *American Economic Review*, 103, no. 3 (May 2013): 611–16, doi: https://doi.org/10.1257/aer.103.3.611.

11. Danial Kahneman, and Amos Tversky, eds. *Choices, Values and Frames* (Cambridge: Cambridge University Press, 2000).

12. W. H. Desvousges, F. Reed Johnson, R. W. Dunford, Kevin J. Boyle, Sara P. Hudson, and K. Nicole Wilson, "Measuring Nonuse Damages Using Contingent Valuation: An Experimental Evaluation of Accuracy, 2nd ed.," *RTI International*, September 28, 2010, doi: https://doi.org/10.3768/rtipress.2009.bk.0001.1009.

13. Jonathan Z. Berman, Alixandra Barasch, Emma E. Levine, and Deborah A. Small, "Impediments to Effective Altruism: The Role of Subjective Preferences in Charitable Giving," *Psychological Science* 29, no. 5 (1 May 2018): 834–44, doi: https://doi.org/10.1177/0956797617747648.

14. Giving What We Can, "Maximise Your Charitable Input," *Giving What We Can* (Website), n.d., accessed July 2, 2022, https://www.givingwhatwecan.org/; 80,000 Hours, "You Have 80,000 Hours in Your Career," *80,000 Hours* (Website), 2021, accessed July 2, 2022, https://80000hours.org/; The Life You Can Save, "Not All Charities Are Equal: Some Charities Can be Hundreds or Even Thousands of Times More Impactful Than Others," *The Life You Can Save* (Website), 2022, accessed July 2, 2022, https://www.thelifeyoucansave.org/; Animal Charity Evaluators, "Learn How You Can Be a More Effective Advocate," Animal Charity Evaluators (Website), 2022, accessed July 2, 2022, https://animalcharityevaluators.org/.

15. See Rosalind Hursthouse, *On Virtue Ethics* (Oxford: Oxford University Press, 1999), p. 28.

16. A challenge raised by Bernard Williams, in his "Replies," in *World, Mind and Ethics: Essays on the Ethical Philosophy of Bernard Williams*, ed. J. Altham and R. Harrison, 185–224 (Cambridge: Cambridge University Press, 1995).

17. For a more extensive argument for this point, see: Shannon Vallor, *Technology and the Virtues: A Philosophical Guide to a Future Worth Wanting* (Oxford: Oxford University Press, 2016), Chapter 1.

18. Aristotle, *Nicomachean Ethics*, trans. C.D.C. Reeve (Indianapolis: Hackett Publishing, 2014), Book II, Chapters 6–9.

19. See: Philippa Foot, *Natural Goodness* (Oxford: Oxford University Press, 2001).

20. Gili Kliger, "Humanism and the Ends of Empire, 1945–1960," *Modern Intellectual History*, 15, no. 3 (2018): 773–800, doi: https://doi.org/10.1017/S1479244317000282.

21. Aristotle, *Politics*, trans. C.D.C. Reeve (Indianapolis: Hackett Publishing, 2017).

22. It is important to note that according to Jewish belief, 'There is no new under the sun' (Ecclesiastes 1:9) as stated by King Solomon; thus, whatever new technology is emerging, its equivalent, in one form or another, was already in existence in the past. An example for such an analogy is the 'Golem,' created by Jewish Kabbalists (the

sage of the Jewish mysticism) across the generations and originating in the first man, and the Talmud. For a thorough discussion of this example, see Goltz, Nachshon, John Zeleznikow, and Tracey Dowdeswell. "From the Tree of Knowledge and the Golem of Prague to Kosher Autonomous Cars: The Ethics of Artificial Intelligence Through Jewish Eyes." *Oxford Journal of Law & Religion,* 9 (2020): 132–156. doi: https://doi.org/10.1093/ojlr/rwaa015.

23. Pan-Indianism is the assumption that all Indigenous thought and culture are similar enough to attribute similar concepts and beliefs to all Indigenous populations. Differences exist, not just between tribes and how they relate to different federal governments (e.g., how Indigenous peoples in United States, Canada, Australia, etc. relate to their respective federal governments), but between Indigenous peoples colonized by the same federal governments (*e.g.,* differences exist between the Mvskoke, Diné, Luiseño, Lenape, etc. in the United States). In North American alone there are more than 1,200 First Nation and Native American communities. The worldviews of these peoples may share some similar themes, but in no way should it be assumed that an individual can speak for Indigenous peoples generally. See: Åke Hultkrantz, *Native Religions of North America* (New York: Harper & Row, 1987), 11, 27; and D. McPherson, and J. D. Rabb, *Indian from the Inside: Native American Philosophy and Cultural Renewal* (Jefferson NC: McFarland and Company, 2011), 12.

24. For example, consider that Thomas Hobbes and John Locke both considered Indigenous nations as real-world examples of the state of nature. See: Thomas Hobbes, *Leviathan* (London, J. M. Dent & Sons, 1931); John Locke, *Two Treatises of Government* (London, 1713).

25. Indigenous concepts and ideas have to be explained in English and explained in reference to western ideas that are taken to be foundational, agreed upon, or otherwise unquestionable. Doing this not only dilutes or changes the ideas, but the expectation that western ideas take priority, or that ideas have to be put in western terms, puts a greater intellectual burden on Indigenous peoples.

26. Since we are both enrolled members of Muscogee Nation, our explanation and understanding of Indigenous ethics will likely be rooted in Mvskoke concepts and Mvskoke ways of life. Although we draw from other sources, these concepts and explanations are not intended to be generalized to all other Indigenous groups. Though similarities may exist, this is just an example of what an account of Indigenous ethics may look like.

27. Viola F Cordova. *How It Is: The Native American Philosophy of V.F. Cordova,* ed. Kathleen Dean Moore, Kurt Peters, Ted Jojola, Amber Lacy, with Foreword by Linda Hogan (Tucson: University of Arizona Press, 2007), 253–4.

28. Cordova, *How It Is,* 253–4.

29. Brian Burkhart, *Indigenizing Philosophy through the Land: A Trickster Methodology for Decolonizing Environmental Ethics and Indigenous Futures* (East Lansing: Michigan State University Press, 2019), 201.

30. Jean Chaudhuri, and Joyotpaul Chaudhuri, *A Sacred Path: The Way of the Muscogee Creeks* (Los Angeles: UCLA American Indian Studies Center, 2001).

31. Joseph Len Miller, "What Do We Need to Know to Live in Harmony with Our Surroundings?" in *Ways of Being in the World: An Introduction to Indigenous Philosophy,* ed. A. Sullivan-Clarke (Peterborough: Broadview Press, forthcoming).

32. Chaudhuri, *A Sacred Path.*

33. See e.g.: Mayi Kuwayu, "Indigenous Data Sovereignty Principles," *Mayi Kuwayu: The National Study of Aboriginal & Torres Strait Islander Wellbeing,* 2022, accessed July 1, 2022, https://mkstudy.com.au/indigenousdatasovereigntyprinciples/; United States Indigenous Data Sovereignty Network (USIDSN), "About the US Indigenous Data Sovereignty Network," *USIDSN,* 2022, accessed July 1, 2022, https://usindigenousdata.org/about-us, which states that the USIDSN "helps ensure that data for and about

Indigenous nations and peoples in the US... are utilized to advance Indigenous aspirations for collective and individual wellbeing"; IndigiData, "Empowering Indigenous Data Leaders," *IndigiData: Indigenous Data Science Education*, 2022, accessed July 1, 2022, https://indigidata.nativebio.org/, which hold workshops for Indigenous data scientists; and S. Imbler, "Training the Next Generation of Indigenous Data Scientists," *New York Times*, June 29, 2021, accessed July 1, 2022, https://www.nytimes.com/2021/06/29/science/indigenous-data-microbiome-science.html.

34. Tahu Kukutai and John Taylor, *Indigenous Data Sovereignty: Toward an Agenda* (Canberra: Australian National University Press, 2016), 2.

35. Stephanie Russo Carroll, Desi Rodriguez-Lonebear, and Andrew Martinez, "Indigenous Data Governance: Strategies from United States Native Nations," *Data Science Journal*, 18, no. 1 (2019): 31, doi: http://doi.org/10.5334/dsj-2019-031.

36. Stephanie Russo Carroll, Ibrahim Garba, Oscar L. Figueroa-Rodriguez, Jarita Holbrook, Raymond Lovett, Simeon Materechera, Mark Parsons, Kay Raseroka, Desi Rodriguez-Lonebear, Robyn Rowe, *et al.*, "The CARE Principles for Indigenous Data Governance," *Data Science Journal*, 19, no. 1 (2020): 43, doi: http://doi.org/10.5334/dsj-2020-043, 43; See also University of Toronto Libraries, "Research Guides: Indigenous Studies: Indigenous Data Sovereignty," *University of Toronto Libraries*, 2022, accessed July 1, 2022, https://guides.library.utoronto.ca/indigenousstudies/datasovereignty; Native Nations Institute, "Indigenous Data Sovereignty and Governance," *Native Nations Institute, University of Arizona*, 2022, accessed July 1, 2022, https://nni.arizona.edu/programs-projects/policy-analysis-research/indigenous-data-sovereignty-and-governance.

3 Research Ethics and the Scientific Method

Brian Wansink refuses to let failure be an option. If he has cool data, he keeps going at it until he finds something, then he publishes, publishes, publishes.

~ (Andrew Gelman, Statistician)[1]

'One Simple Trick': The Cornell Food & Brand Lab

Did you know that you are 80% more likely to choose a salad if you sit by the window in your favorite restaurant?[2] Or that if you sit near the bar – in dim lighting and with loud music playing in the background – you will consume more calories?[3] Did you know that people who keep their cereal boxes out on the counter weigh 21 lbs more on average than those who keep them hidden in the cupboard?[4] Or that branding apples with popular cartoon characters, like Elmo, makes children more likely to choose one over a cookie with their lunch?[5] Or that men eat more when women are watching them – but that women eat *less* when men are watching *them*?[6] Or that striking a 'power pose' has a positive impact on job interviews, negotiations, and other performances – especially for those with lower social status and fewer resources?[7]

If you have answered 'no' to any or all of these questions, then you can congratulate yourself, for you are correct. The claims made by researchers in the above studies – all once prominently promoted in the media – were not reproducible and have since been retracted. Amy Cuddy's work on power poses was the subject of the second most watched TED Talk ever, and even became part of our received cultural folk wisdom, before its retraction.[8] The other claims, too, made their way into the folk wisdom of those interested in the latest diet and health news – including those in charge of making decisions about public school lunch programs. They, too, have all been retracted after findings of research misconduct.[9]

These studies were the product of Brian Wansink of Cornell University, where he studied the psychology of eating at his Cornell Food & Brand Lab.[10] Wansink founded the Food & Brand Lab at the University of Illinois in 1997 and moved it to the Ivy League in 2005.[11] The Food & Brand Lab received much of its funding from food companies.[12] Wansink's experiments were not only well-funded but popular and highly publicized.[13] His book, *Mindless*

DOI: 10.1201/9781003293125-3

Eating: Why We Eat More Than We Think, hit the New York Times bestseller list in 2006.[14] His philosophy was quite different from the prevailing wisdom at the time: Wansink believed that, instead of teaching people about the benefits of good dietary choices and the dangers of poor ones, he could get people to adopt simple tricks and habits that would 'nudge' them toward eating better without their thinking too much, or having to be in any way rational, about their choices. "There are a million nutritionists out there that tell you to eat an apple instead of a Snickers bar," he told Kiera Butler of *Mother Jones* in 2015, "If we really want to eat better... we have to trick our brains into making the right choices."[15]

However, other scientists began to raise concerns over Wansink's research methods, including "data inconsistencies, mathematical impossibilities, errors, duplications, exaggerations, eyebrow-raising interpretations, and instances of self-plagiarism – in 50 of his studies"[16] – many of which have since been retracted.[17] These include several papers describing how attractive presentations of healthy foods in school cafeterias promote students' choosing more fruits and vegetables.[18] Programs based on Wansink's retracted publications have been adopted in 30,000 U.S. schools, which have attracted millions of dollars in government funding for the Smarter Lunchrooms Movement.[19] The programs mainly involved giving healthy food snappy and attractive names and colorful branding, like "orange squeezers," "monkey phones (bananas)," "snappy apples," "cool-as-a-cucumber slices," and "sweetie pie sweet potatoes."[20]

The cracks in the research were visible early on, but they came to a head because of a blog posting by Wansink himself – in what must be one of the more consequential acts of hubris in the history of science. In the blog, Wansink discusses an original data set collected over a few weeks of observation at a single pizza restaurant in upstate New York; he notes that that the original research plan did not pan out, so he was looking to mine the data for some "cool" new research results.[21] He then strongly criticized his (paid) post-doc for refusing to work with the data, while a visiting (unpaid) post-doc from Turkey, mined the data and ended up publishing five different papers – the now infamous "pizza papers."[22]

Wansink congratulated the Turkish researcher's inventiveness and initiative in making so much of the data while stating, "In comparison, the post-doc left after a year (and also left academia) with 1/4 as much published (per month) as the Turkish woman. I think the person was also resentful of the Turkish woman."[23] Wansink concluded with the following parting shot to his former employee, stating, "While most of us will never remember what we read or posted on Twitter or Facebook yesterday... this Turkish woman's resume will always have the five papers below."[24] Except she didn't, because they were retracted for research misconduct.

Tim van der Zee of Leiden University in The Netherlands, was one of the first scientists to read Wansink's blog post and call out possible misconduct in the pizza papers.[25] The studies in the retracted pizza papers took place at a buffet restaurant, named *Aiello's Italian Restaurant,* about 30 miles from

Cornell. The sample consisted of about 130 adults who had eaten at the restaurant over a two-week period.[26] The authors noted that it was a breach of publication ethics not to state in any of the papers that the data all came from the same, previously-published, field study.[27] They were also denied access to the original data when they requested it from Wansink.[28] They noted that the sample sizes were inconsistent between the papers, suggesting that some participants were *included* in some papers, but *excluded* in others.[29] Van der Zee also noted several other errors in the paper:

> The types of errors include: impossible sample sizes within and between articles, incorrectly calculated and/or reported test statistics and degrees of freedom, and a large number of impossible means and standard deviations. In total, we identified approximately 150 inconsistencies and impossibilities in these four papers. Taken together, these problems make it difficult to have confidence in the authors' conclusions.[30]

At first, Wansink dismissed the errors as minor and the criticisms as "cyberbullying,"[31] but the calls for a full investigation into his research only grew.

Andrew Gelman, a prominent statistician at Columbia University, then called out Wansink in a scathing blog post of his own.[32] Gelman stated, "What you describe Brian does sound like p-hacking and HARKing. The problem is that you probably would not have done all these sub-group analyses and deep data dives if your original hypothesis had $p < .05$."[33] Here, Gelman is referring to the process of 'hypothesizing after results are known' (HARKing)[34] – in this case, Wansink's original hypothesis seems to have found no support, so the data was simply mined by the Turkish post-doc to see if some plausible associations turned up. Bell recommends that researchers can avoid HARKed false positives by declaring "clearly-motivated hypotheses, alongside falsifiable predictions, before experimentation."[35] This is being accomplished in many fields, including machine learning, through the preregistration of experiments, including hypotheses, data, analysis, and experimental design.[36] The OpenML repository is a good example of the move toward open science.[37]

By p-hacking, Gelman is referring to the discredited practice of massaging data – for example by playing around with sample sizes – to generate a seemingly statistically-significant result where none actually exists.[38] P-hacking, too, compromises the validity of the models because it "invalidates the core assumption of a statistical hypothesis test: the probability that a single result is due to chance."[39] P-hacking can lead us to accept as valid results that are merely fortuitous. P-hacking joins HARKing, data dredging, and reporting marginally significant results as practices that produce invalid models in machine learning as well.[40] The large data sets used in machine learning are particularly liable to produce false positives[41] (see Box 3.1 for the definitions of the main aspects of the scientific method).

Gelman concluded his blog post by stating, "I'm bothered by that last sentence that the resume 'will always have the five papers.' The end state of

Box 3.1
The Scientific Method

Reproducibility: The results obtained in one trial or experiment will be similar when repeated under similar conditions, which need to be documented by researchers in a way that is complete as well as transparent. Also known as replicability and repeatability.

Reliability: The reliability and internal consistency of a measurement or result.

Stability: A measure of reliability. Also known as test/retest reliability. An individual who takes a test several times will give largely the same answers. A system that runs several times under the same conditions will produce largely the same results over time.

Precision: Measurements or trial runs that produce outcomes that closely resemble each other. Low variation.

Accuracy: A measurement of error between the average measurements and the true value.

Validity: The extent to which a purported model or measurement accurately reflects what it claims to.

Hypothesis: A provisional explanation for an observed phenomenon that is capable of being tested and disproved. A null hypothesis is one which posits that there is no relationship between two phenomena, and that the results observed are due to chance. A working hypothesis is one that has been provisionally accepted and is actively being tested.

Type I Error: The null hypothesis is true but is wrongly rejected due to a seemingly statistically significant result. A false positive.

Type II Error: The null hypothesis is false but is wrongly accepted.

Statistical Significance: The null hypothesis warrants rejection because the observed results are unlikely to be due to chance. Represented by a p-value. Historically, a statistically significant p-value has been $p > 0.05$, meaning that the chances that a positive result is due to random chance alone are 5%, although the sufficiency of this p-value has recently been called into question.

p-Hacking: Any practice that manipulates data to derive a statistically significant p-value where one is absent. Results in a type I error.

HARK-ing: Hypothesizing After Experimental Results are Known. The results of an experiment or trial are known, and then a hypothesis is constructed to explain the observed data. A null hypothesis is not formulated and tested against the observed results.

Cherry-Picking: Experiments or trials that produce negative results are not reported, while trials or runs that produce positive results are reported. Results in a type I error: wrongly accepting a null hypothesis that is not true.

research is not the resume. Nor is it the tenured job, the press release, the Ted talk, or the appearances on Oprah and Dr. Oz. Just ask Roy Baumeister or John Bargh"[42] – referring to a scientist (Bargh) who had an influential paper retracted for p-hacking.[43]

The scandal was career ending for Wansink. Other scientists began asking for the original data in the school luncheon studies, but none could not be found.[44] The branding and school lunch paper was then retracted, too.[45] In September 2018, Wansink retired after an investigation at Cornell found that he had indeed committed research misconduct, including data misreporting, missing data, statistical errors, and inappropriate authorship attributions. The year before, their investigation had found 'errors', but no 'misconduct'.[46]

The criticisms of Wansink's research came at a sensitive time for the replication crisis, and Singal states that his was one of the great tragedies of that crisis.[47] Wansink and his lab were prolific publishers of attention-grabbing studies – a practice that often leads to quality control errors such as we saw here.[48] As Signal has stated, "Many of psychology's most exciting "This One Simple Trick Can X"–style findings have turned out to be little more than statistical noise shaped sloppily into something that, in the right light and if you don't look too hard, looks meaningful."[49]

The reproducibility crisis, of course, is about much more than reproducibility. It is about the nature of the scientific method itself (Box 3.1), and what it means to produce theories, models, and – ultimately – knowledge that gives an objectively correct picture of reality. Many results in psychology, medicine,[50] and the social sciences[51] have been found to be not reproducible – and therefore probably invalid as well (Box 3.2).[52]

Box 3.2

Reproducibility Checklist

For all *models* and *algorithms* presented, check if you include:

- A clear description of the mathematical setting, algorithm, and/or model.
- A clear explanation of any assumptions.
- An analysis of the complexity (time, space, sample size) of any algorithm.

For any *theoretical* claim, check if you include:

- A clear statement of the claim.
- A complete proof of the claim.

For all *datasets* used, check if you include:

- The relevant statistics, such as number of examples.
- The details of train/validation/test splits.
- An explanation of any data that were excluded, and all pre-processing steps.
- A link to a downloadable version of the dataset or simulation environment.
- For new data collected, a complete description of the data collection process, such as instructions to annotators and methods for quality control.

For all shared *code* related to this work, check if you include:

- Specification of dependencies.
- Training code.
- Evaluation code.
- (Pre-)trained model(s).
- README file includes a table of results accompanied by the precise command to run to produce those results.

For all reported *experimental results*, check if you include:

- The range of hyper-parameters considered, the method to select the best hyper-parameter configuration, and the specification of all hyper-parameters used to generate results.
- The exact number of training and evaluation runs.
- A clear definition of the specific measure or statistics used to report results.
- A description of results with a central tendency (e.g., mean) and variation (e.g., error bars).
- The average runtime for each result, or estimated energy cost.
- A description of the computing infrastructure used.

Source: Pineau, Joelle. The Machine Learning Reproducibility Checklist (v2.0, Apr.7 2020). www.cs.mcgill.ca/~jpineau/ReproducibilityChecklist-v2.0.pdf.

Therefor, the reproducibility crisis refers to poor methodology, as well as the lack of validity: results produced through unethical methodologies will result in models that are not valid – and therefore do not give us reliable information about the real world. This book will discuss several case studies where this can lead to significant harms – including wrongful convictions, unnecessary detentions, dangerous persons being released under inadequate conditions, and even genocide, ethnic cleansing, and political violence.

Of course, research ethics will only lead to valid and robust results if the field itself creates a culture that values scientific ethics and methodological rigor.[53] Psychologists have found that fostering an ethical research culture helps to ensure not only reproducibility but the actual validity of published results.[54] Good methodology also creates trust among researchers. As Heil notes, "no scientist can reproduce results from every paper they read,"[55] and very few published papers will even see a single attempt at reproduction. The rest we take on *trust*.

Gencoglu makes a point that we will come back to many of the case studies that follow: a rigorous research culture in machine learning needs to "address human needs and psychology in a realistic manner."[56] To do so, "high-level domain experts should be incorporated to the study teams from the beginning,"[57] especially as machine learning forays into fields that have long developed their own expertise – forensic evidence, risk assessments in criminology, biometrics, media effects, and free speech laws, among others.

In the end, there is no 'one simple trick' to ensure that research is giving us valid knowledge and providing an accurate and useful picture of the reality that we are trying to understand and model – just as there are no simple tricks to learning how to eat healthy, deciding what content should be prohibited on social media, or determining guilt and innocence in a criminal trial. In a young and rapidly growing field like machine learning, it will take a culture that values robust methods and the validity of models – one that is reflective about producing knowledge that serves people's needs, and that stands the test of time.

Commentary

Utilitarian Ethics

By Peter Singer and Yip Fai Tse

From a Utilitarian perspective, Wansink's conduct is unethical because it creates a risk for more negative consequences than potential gains. Imagine a scientific field where the majority, or even just a substantial portion, of practitioners are not intellectually honest. Research in that field could not be relied upon.

Wansink seems to have an agenda behind his research: he wanted people to eat a certain (healthy, as he might have believed) way. This might be the reason he only favored the results that support his views. To wish that people eat in a certain way is, of course, not necessarily bad. And it is possible, maybe likely, that his intention was good. But a good intention does not justify being intellectually dishonest.

Holding good intentions, by itself, is not enough in order to act ethically; one must also act in an evidence-based, empirically, and theoretically sound way.[58] A well-intentioned person, finding evidence or arguments against their agenda, needs to re-evaluate, and perhaps, if the reasons are strong enough, reject their agenda.

Ignoring evidence and arguments against one's agenda, may turn good intentions into self-deceiving fantasies; it can also cause serious harm, possibly on a vast scale. In Wansink's case, he risked much more than his own career and the reputation of his field and institution; he also risked giving ill-informed advice on dietary habits, and thereby harming those who followed his advice.

Intellectual honesty is not the only ethical requirement. Researchers, especially those working on projects that can potentially harm the lives of sentient beings, are ethically responsible for the foreseeable impacts of their research. For example, the impact of research in biology can be significant, because it often has major consequences on many humans and nonhuman animals.

The recent concern that CRISPR technology can be used to enable terrorists to modify viruses for attack purposes is just one example of how biotechnology can cause immense impacts.

Data science has at least as much expected impact as biology. It is important that researchers think carefully about the ethical implications of their work before publishing, or better still, before even conducting their research in certain fields.

Notes

1. Andrew Gelman, "Hark, Hark! The p-value at Heaven's Gate Sings." *Statistical Modeling, Causal Inference, and Social Science* (Blog). December 15, 2016, accessed December 19, 2021, https://statmodeling.stat.columbia.edu/2016/12/15/hark-hark-p-value-heavens-gate-sings/.
2. Jesse Singal, "A Popular Diet-Science Lab Has Been Publishing Really Shoddy Research," *The Cut*, February 8, 2017, https://www.thecut.com/2017/02/cornells-food-and-brand-lab-has-a-major-problem.html.
3. Singal, "Popular Diet-Science."
4. Kiera Butler, "This Fast-Food-Loving, Organics-Hating Ivy League Prof Will Trick You into Eating Better," *Mother Jones*, last Modified September 20, 2018, https://www.motherjones.com/environment/2015/03/brian-wansink-cornell-junk-food-health/.
5. B. Wansink, David R. Just, and Collin R. Payne, "Notice of Retraction and Replacement. Wansink B, Just DR, Payne CR. Can Branding Improve School Lunches?" *Archives of Pediatric and Adolescent Medicine,* 166, no. 10 (2012): 967–968, doi: https://doi.org/10.1001/archpediatrics.2012.999
6. Steven Salzberg, "Cornell's Food Lab is Cooking Up Fake News," *Forbes*, October 2, 2017, https://www.forbes.com/sites/stevensalzberg/2017/10/02/cornells-food-lab-is-cooking-up-fake-news/?sh=24d020be25cd; Gelman, "Hark, Hark!"
7. Joseph Cesario, Kai J. Jones, and Dana R. Carney, "CRSP Special Issue on Power Poses: What Was the Point and What Did We Learn?" *Comprehensive Results in Social Psychology,* 2, no. 1 (2017):1–5, doi: https://doi.org/10.1080/23743603.2017.1309876.
8. Cesario, "Power Poses."
9. B. Wansink, David R. Just, and Collin R. Payne, "Notice of Retraction and Replacement. Wansink B, Just DR, Payne CR. Can Branding Improve School Lunches?" *Archives of Pediatric and Adolescent Medicine,* 166, no. 10 (2012):967–968, doi: https://doi.org/10.1001/archpediatrics.2012.999.
10. Stephanie M. Lee, "Emails Show How an Ivy League Prof Tried to Do Damage Control for His Bogus Food Science," *Buzz Feed News*, September 27, 2017, https://www.buzzfeednews.com/article/stephaniemlee/brian-wansink-cornell-smarter-lunchrooms-flawed-data.
11. Kiera Butler, "This Fast-Food-Loving, Organics-Hating Ivy League Prof Will Trick You into Eating Better," *Mother Jones*, last modified September 20, 2018, https://www.motherjones.com/environment/2015/03/brian-wansink-cornell-junk-food-health/.
12. Butler, "Fast-Food-Loving."
13. Lee, "Emails Show."
14. Butler, "Fast-Food-Loving."
15. Butler, "Fast-Food-Loving."
16. Lee, "Emails Show."
17. Lee, "Emails Show."
18. Lee, "Emails Show."
19. Lee, "Emails Show." Proponents of 'nudging' students to make healthy choices in school lunches counter that there is support for these techniques in studies done by other researchers.
20. Lee, "Emails Show."
21. Gelman, "Hark, Hark!"

22. Singal, "Popular Diet-Science."
23. Singal, "Popular Diet-Science."
24. Singal, "Popular Diet-Science."
25. Tim Van der Zee, Jordan Anaya, and Nicholas J. L. Brown, "Statistical Heartburn: An Attempt to Digest Four Pizza Publications from the Cornell Food and Brand Lab," *BMC Nutrition,* 3, no. 54 (2017), doi: https://doi.org/10.1186/s40795-017-0167-x. The original blog post, which has since been taken down, is referenced here as: Brian Wansink, "The Grad Student Who Never Said 'No,'" *Healthier & Happier,* 2016, https://web-beta.archive.org/web/20170312041524/. Accessed 24 June 2017.
26. Van der Zee, "Statistical Heartburn."
27. Van der Zee, "Statistical Heartburn."
28. Van der Zee, "Statistical Heartburn."
29. Van der Zee, "Statistical Heartburn."
30. Van der Zee, "Statistical Heartburn."
31. Lee, "Emails Show."
32. Gelman, "Hark, Hark!"
33. Gelman, "Hark, Hark!"
34. Samuel J. Bell, and Onno P. Kampman, "Perspectives on Machine Learning from Psychology's Reproducibility Crisis, v. 1," *arXiv.org,* April 18, 2021, last modified April 23, 2021: arXiv:2104.08878, 2.
35. Bell, "Perspectives on Machine Learning," 2.
36. Brian A. Nosek, Charles R. Ebersole, Alexander C. DeHaven, and David T. Mellor, "The Preregistration Revolution," *Proceedings of the National Academy of Sciences,* 115, no. 11 (2018): 2600–2606, doi: https://doi.org/10.1073/pnas.1708274114.
37. Open ML, Website, 2022, accessed July 1, 2022, https://www.openml.org/.
38. Bell, "Perspectives on Machine Learning," 2. Bell states that p-hacking "invalidates the core assumption of a statistical hypothesis test: the probability that a single result is due to chance."
39. Bell, "Perspectives on Machine Learning," 2.
40. Gencoglu, "HARK Side of Deep Learning," 2.
41. Bell, "Perspectives on Machine Learning," 2.
42. Gelman, "Hark, Hark!"
43. Andrew Gelman, "Priming Effects Replicate Just Fine, Thanks," *Statistical Modeling, Causal Inference, and Social Science* (blog), February 12, 2016, https://statmodeling.stat.columbia.edu/2016/02/12/priming-effects-replicate-just-fine-thanks/. The much-cited retracted paper claimed that students walked slower when primed with 'elderly-related' concepts, like 'bingo' and 'Florida'. See also: Andrew Gelman, "It Comes Down to Reality and It's Fine with Me Because I've Let it Slide," *Statistical Modeling, Causal Inference, and Social Science* (blog), June 23, 2016, https://statmodeling.stat.columbia.edu/2016/06/23/it-comes-down-to-reality-and-its-fine-with-me-cause-ive-let-it-slide/. Here, Gelman criticizes Baumeister for defending creativity and inventiveness when it comes to finding statistically significant result in very small sample sizes, which Gelman sees as a misguided defence of p-hacking.
44. Lee, "Emails Show."
45. Lee, "Emails Show."
46. Butler, "Fast-Food-Loving."
47. Singal, "Popular Diet-Science."
48. Singal, "Popular Diet-Science."
49. Singal, "Popular Diet-Science."
50. Andrew L. Beam, Arjun K. Manrai, and Marzyeh Ghassemi, "Challenges to the Reproducibility of Machine Learning Models in Health Care," *JAMA,* 323, no. 4. (2020): 305, doi: https://doi.org/10.1001/jama.2019.20866.

51. Matthew Hutson, "Artificial Intelligence Faces Reproducibility Crisis," *Science,* 359, no. 6377 (2018): 725–726.
52. One recent 8-year review, for example, found that only about ¼ of cancer research papers could be reproduced. See *e.g.*: Tara Haelle, "A Massive 8-Year Effort Finds that Much Cancer Research Can't Be Replicated," *Science News*, December 7, 2021, https://www.sciencenews.org/article/cancer-biology-studies-research-replication-reproducibility.
53. Bell, "Perspectives on Machine Learning," 4.
54. Hutson, "Artificial Intelligence," 725–6.
55. Heil, "Reproducibility Standards."
56. Gencoglu, "HARK Side of Deep Learning," 6.
57. Gencoglu, "HARK Side of Deep Learning," 6.
58. This is the core principle of the Effective Altruism movement. See: Peter Singer, *The Most Good You Can Do: How Effective Altruism s Changing Ideas About Living Ethically* (Yale University Press, New Haven, 2015).

4 Machine Models in Court

Scientific conclusions are subject to perpetual revision. Law, on the other hand, must resolve disputes finally and quickly. The scientific project is advanced by broad and wide-ranging consideration of a multitude of hypotheses, for those that are incorrect will eventually be shown to be so, and that in itself is an advance. Conjectures that are probably wrong are of little use, however, in the project of reaching a quick, final, and binding legal judgment – often of great consequence – about a particular set of events in the past.

~ (Justice Blackmun, *Daubert* v. *Merrell Dow Pharmaceuticals*, U.S. Supreme Court, 1993)[1]

The Trials of Oral Nicholas Hillary

On October 24, 2011, a strange and terrible homicide took place in Potsdam, New York – a small town next to the St. Lawrence River and very close to the border with the Canadian province of Ontario. A 12-year-old boy named Garrett Phillips was strangled to death in his bedroom around 5 pm in the afternoon, a short time after he returned home from school.[2] Neighbors heard a commotion and phoned 911. The blinds on Garrett's second-story bedroom window were bent outward, leading investigators to suspect that the killer jumped out that way and ran off.[3]

Police quickly settled on Oral Nicholas (Nick) Hillary as their lead suspect in the case, and for reasons that appear wholly specious. Hillary was the soccer coach for the varsity men's team at Clarkson University and was successful and very popular – almost everyone in Potsdam knew who he was.[4] He had recently been in a romantic relationship with Tandy Cyrus, Garrett's mother. He was also one of a small number of African Americans living in Potsdam,[5] and his relationship with Tandy had caused some shockwaves in the community. They had recently split up because Tandy's two sons did not get along with Nick, which had caused problems within the family.[6]

What's more, Nick and Tandy were both in relationships with other people when their own began. At the time, Tandy was seeing John Jones – a sheriff in Potsdam. John Jones was quite upset that Nick was a factor in his break-up with Tandy, and he went to Nick's house to confront and, possibly, even threaten him.[7] There were clearly tensions in the community, and a general

DOI: 10.1201/9781003293125-4

feeling that John was not only put out over the loss of his girlfriend, but to have lost out to a rival who was an African American and very successful.[8]

The police detained Hillary as a suspect very quickly, in what seems to be a clear case of tunnel vision, caused by racial and personal animus.[9] He was detained and interrogated for several hours and was even strip searched to see if he had any injuries consistent with a jump from a second-story window.[10] He did not. With no evidence linking him to Garrett's murder, he was released and would later file a civil rights lawsuit against the police.[11]

This was the beginning of Hillary's legal troubles rather than the end. The attorney defending against the civil rights lawsuit developed a strategy of proving Nick did in fact commit the murder – and he used the testimony Hillary provided during his deposition against him.[12] Despite the very shaky evidence, District Attorney Mary Rain obtained an indictment for second-degree murder against Hillary on May 12, 2014.[13] This indictment was dismissed in October 2014 due to prosecutorial misconduct on Rain's part.[14] On February 2, 2015, Rain convened a second grand jury and obtained another indictment for murder against Hillary.[15]

The community was in turmoil as the case dragged on with no clear solution. There were widespread rumours that there was strong DNA evidence against Hillary and that it was being suppressed due to 'technicalities'.[16]

The 'Most Dangerous Prosecutor in New York'

That DNA evidence was collected and interpreted in a rarified climate of racial animus against Hillary and clear evidence of prosecutorial misconduct on the part of DA Mary Rain. Rain originally ran for the office on a platform of solving Garrett's homicide, leading to criticisms that she was using the murder for political gain – she often appeared alongside Tandy Cyrus at campaign events.[17] Rain quickly used her new position to harass those officials who had fired her for incompetence when she was a public defender.[18] She left the office in 2017 under a cloud of suspicion without seeking re-election.[19] She would later be suspended from practicing law for two years – a very rare occurrence, and a testament to the severity and persistence of her misconduct as a prosecutor.[20]

Indeed, Rain's misconduct during her tenure as a DA was notorious.[21] In the case against Hillary, it included several unlawful acts of non-disclosure.[22] She suppressed the fact that a witness had reported seeing John Jones entering Garrett's apartment near the time he was killed.[23] Tandy Cyrus had laid a complaint against Jones in January of 2011, stating that he was acting in ways that caused her to fear for her safety and the safety of her children, including Jones going into her apartment unannounced and uninvited despite being repeatedly told not to.[24]

Although he was a reasonable suspect in the case, Jones gave Rain an alibi that she readily accepted. Rain justified her suppression of the evidence by claiming that the witness's statement "did not go with the [prosecution's] theory of the case," and so there was no reason to disclose it to the defence.[25] This is probably the most culpable and unethical reason possible for a prosecutor to fail to disclose exculpatory evidence – particularly when the suspect in question is a local sheriff who was actively involved in the murder investigation.[26]

The first grand jury case against Hillary was dismissed due to Rain's unethical conduct. Judge Richards ruled that Rain tainted the proceedings, including by badgering Hillary's 17-year-old daughter to disclose communications protected by the attorney-client privilege.[27] At the same time, the FBI was investigating Rain for contacting inmates without their lawyer's consent to convince them to testify against other inmates.[28] Jailhouse informants who have been pressured to testify while being denied their right to counsel produce highly unreliable evidence, and this has been shown to be a significant contributing cause of wrongful convictions.[29]

The DNA Evidence

Hillary's trial for the murder of Garrett Phillips was held before a judge alone in St. Lawrence County. Hillary may have chosen a bench trial because he thought – probably correctly – that a local jury might not be fair to him. There was little physical evidence available at the trial. Four latent fingerprints were found on and around the second-story window where the perpetrator was believed to have escaped.[30] The fingerprints were not Hillary's and they have never been matched to anyone known to be connected to the case, or anyone in the New York State SAFIS database.[31]

There were also trace amounts of DNA evidence, which ended up being dispositive of the case. The DNA profile was developed from fingernail scrapings collected at Garrett's autopsy. Investigators developed a theory that Garrett may have struggled with and scratched his attacker before his death.[32] However, DNA could only be recovered in trace amounts, suggesting that there may have been significant degradation of the DNA, or that it could have been part of background or investigator-mediated contamination.[33] Because the DNA was low-copy number, it fell below the recommended threshold for analysis.[34] This, in turn, makes the interpretation of the DNA profile much more difficult and calls into question the probative value of the evidence.[35]

Interpreting degraded, low-copy number, and mixed DNA profiles has proven to be a challenge for forensic scientists and courts alike.[36] Due to the complexity of calculating the probability that a given individual's DNA profile is included in the sample taken from the crime scene, several algorithms have been developed to estimate probabilities of inclusion and likelihood ratios.[37] Traditional methods of estimating probabilities for low-copy number and mixed samples assign an equal probability to all genotypes, which limits the probative value of this evidence. DNA interpretation algorithms assign statistical weights to different genotypes – including the probability that certain alleles might 'drop out' and not appear in the sample, or that an artifact might appear as an allele and thus 'drop in' when it is not in fact included.[38] The DNA profile developed from under Garrett's fingernails was a partial profile, meaning that several alleles had dropped out, and could not be detected in the electropherogram. These algorithms therefore do a much better job of estimating whether a given individual is or is not included in a complex sample taken from a crime scene.[39] Two of the most popular commercially available DNA mixture interpretation software models are TrueAllele[40] and STRmix.[41] Both were used in Hillary's case, and both came to different conclusions on whether his DNA was included in the profile developed from Garrett's fingernail scrapings.

John Buckleton, a leading forensic geneticist who played a key role in developing STRmix, states that DNA mixture algorithms can analyze complex DNA samples with greater speed and accuracy.[42] DNA interpretation software generally works by using Monte Carlo Markov Chain (MCMC) methods to resolve mixtures and develop conditional probabilities of inclusion.[43] MCMC methods have long been used in machine learning models – as they have long been used in many fields, including physics, econometrics, and computer science – to solve high-dimensional problems.[44] It is important to incorporate high-level domain knowledge into the MCMC model to generate a suitable list of candidate hypotheses.[45]

Candidate hypotheses in this case would include whether Hillary's DNA is included in the sample;[46] whether he is excluded as being a contributor to the sample;[47] whether the DNA sample came to be under Garrett's fingernails via background contamination and is not probative evidence in his homicide (it is not uncommon to find other people's DNA under our fingernails in trace amounts due to contamination from everyday activities);[48] and whether the DNA was introduced after the crime through some form of investigator-mediated contamination.[49]

For DNA interpretation, this domain knowledge also includes the rules for finding evidence to be admissible in court (Box 4.1), as well as the *SWGDAM Guidelines* for validating probabilistic genotyping systems.[50] Prior to use, a forensic DNA analyst needs to interpret the peaks, estimate drop-in and drop-out, and call which are in fact alleles.[51] The software must first be internally validated by the lab under conditions that are similar to those of the crime scene sample.[52] These basic quality controls were not used in Hillary's case. Judge Felix Catena held a *Frye* hearing to determine if the STRmix evidence would be admissible given that it was derived from low-copy number DNA and consisted of only a partial DNA profile.[53] The *Frye* test is one of the two main standards courts use in the United States to determine if expert evidence is admissible in court (Box 4.1). The essence of the *Frye* test is that expert evidence will be admitted if the science on which the opinion is based is generally considered reliable in that scientific community.[54] The other key test of admissibility, set out in *Daubert*,[55] looks not only to whether the technique or theory is accepted, but whether it can be and has been tested, whether there is a known error rate for the procedure, whether there are quality control and other standards governing the procedure, and whether it has been subject to peer review (Box 4.2).

Box 4.1
___

Basic Rules of Evidence

The rules of evidence are voluminous, and each jurisdiction will have its own variations. Evidence will generally be admissible in court if it is:

Relevant: Evidence is relevant if it helps the court to answer a question that is in dispute or tends to prove or disprove a fact that is of importance. The *probative value* of the evidence refers to how much a piece of evidence tends to prove or disprove the fact in dispute.

Reliable: Evidence that is unreliable, or that will tend to mislead a trier of fact (a jury, or a judge sitting alone) – perhaps because it is too complicated to be communicated adequately – will not be admitted. Evidence that comes from first-hand knowledge, or that was gathered in accordance with quality control procedures or by an accredited laboratory facility, will often be considered more reliable. The *weight* to be given to the evidence by the trier of fact will often depend on how reliable the evidence is.

Necessary: The evidence is necessary to prove or disprove a fact or issue in dispute. Evidence will be unnecessary if it duplicates other evidence presented. On the other hand, evidence might be necessary if there is no other way for a party to put that evidence before a court, which will be considered by the court when assessing whether the evidence is reliable and fair. Evidence gathered by a machine learning system in the absence of any human operator may therefore be necessary.

Not Excluded: There are many exclusionary rules of evidence that might require the court to exclude evidence that would otherwise be admissible. For example, the rules of hearsay, or privileged communications, can result in the exclusion of even reliable and probative evidence.

Fair: The admission of the evidence will often be considered fair if its probative value outweighs the unfair *prejudice* it may cause to a party. Many jurisdictions will also have their own constitutional laws that govern police searches, and that help courts determine when the evidence has been gathered in an unreasonable way, such as the Fourth Amendment of the U.S. Constitution.

In this case, the forensic DNA evidence would not have met either the *Frye* or *Daubert* standards. Initial results from the DNA mixture evidence were analyzed by TrueAllele, but due to the poor quality of the DNA profile, no result could be obtained: Hillary could neither be included nor excluded from the sample.[56] TrueAllele concluded at the time that they could find "no statistical support" that Hillary had contributed to the DNA mixture taken from under Garrett's fingernails.[57] TrueAllele states that it examined over 100 items of interest in the case, "and showed that Hillary was not connected to the DNA evidence in this case."[58]

Box 4.2

Admissibility of Scientific and Expert Evidence in the United States

Frye **Test:** Also called the general acceptance test. A court will admit scientific or expert evidence if it is generally accepted in the relevant scientific community. This test is set out in *Frye v. United States*, 293 F. 1013 (D.C. Cir. 1923) in a case that excluded polygraph evidence because it was not generally accepted as being reliable. Some U.S. jurisdictions use this test, such as Washington, California, Illinois, Minnesota, New York, New Jersey, and Pennsylvania. Most other states have adopted the *Daubert* test.

Daubert **Test:** This standard of expert evidence is set out in *Daubert v. Merrell Dow Pharmaceutical Inc.*, U.S. 579 (1993) and interprets Rule 702 of the Federal Rules of Evidence to mean that judges must exercise a gatekeeping function to ensure that scientific and expert evidence is relevant and reliable. Scientific knowledge is one that is gathered in accordance with the scientific method, and this depends on much more than whether it is generally accepted. The court may also look to see if the methods have been tested and validated, whether there is a known error rate, whether it has been peer-reviewed, whether it has been produced specifically for the present case or has become accepted and used in a variety of fields and applications.

Rule 702. Testimony by Expert Witnesses: A witness who is qualified as an expert by knowledge, skill, experience, training, or education may testify in the form of an opinion or otherwise if:

a   The expert's scientific, technical, or other specialized knowledge will help the trier of fact to understand the evidence or to determine a fact in issue;
b   The testimony is based on sufficient facts or data;
c   The testimony is the product of reliable principles and methods; and
d   The expert has reliably applied the principles and methods to the facts of the case.

Source: U.S. Federal Rules of Evidence, as amended Apr. 17, 2000, effective Dec. 1, 2000; and as amended Apr. 26, 2011, effective Dec. 1, 2011.

William Fitzpatrick, a DA from Onondaga County, then contacted John Buckleton to see if his DNA algorithm, STRmix, could obtain a different result.[59] They looked at the DNA found under Garrett's fingernails, and the partial profile was found to 'include' Nick Hillary.[60]

The New York State Crime Laboratory first amplified the DNA in the tissue sample via a PCR (Polymerase Chain Reaction), but appears to have run the reaction more times than recommended in an attempt to pull up more of the trace DNA[61] – this increases stochastic effects and can lead to 'drop-in' where noise appears as an allele in the DNA profile.[62] There was also bias in the way the analyst initially assessed the partial profile, as he not only knew that Hillary was a suspect but called the alleles in the profile *by referencing Hilary's own profile.* Lyons states that "his work notes suggest he reviewed Hillary's DNA profile while trying to match it to the evidence."[63] This contradicts best practices in forensic science whereby an analyst should be blind as to who potential suspects are, and what a suspect's DNA profile is, so as to minimize cognitive and contextual biases in their analysis.[64]

The lab chose an arbitrary cutoff of 50 rfu to call the alleles.[65] This seemed to have little foundation other than to help exclude some alleles present in Hillary's profile, and that would have required the analyst to conclude that he was not a contributor to the sample. Mark Perlin of TrueAllele testified that there were peaks just below this threshold that did exclude Hillary, and therefore the DNA sample was exculpatory.[66]

Nor were proper validation studies conducted by the lab, as recommended by STRmix.[67] When Buckleton ran the STRmix algorithm, he was given only the electropherogram – prepared in a biased fashion by the analyst – and so was forced to "pick and choose data from different 'reliable sources' and input parametrics into the program in such a way that he believed the system would tolerate."[68] The New York State Crime Laboratory was not authorized to use STRmix without validation studies, and as Buckleton himself had recommended.[69] Nor was the lab following SWGDAM's own *Guidelines*, which required thorough internal validation by the lab of complex, low-copy number, and mixed samples.[70]

For these reasons, Justice Catena excluded the DNA evidence.[71] This resulted in Hillary's acquittal, as there was little other evidence against him.[72]

Mary Rain stated after the acquittal that she was 100% certain of Hillary's guilt, despite the lack of evidence confirming this – and the numbers of other suspects that had been put forth by town gossip and the reporters covering the case.[73] Rain stated that there would be no search for anyone else, for no one else could have committed the crime.[74] The new DA, Gary Pasqua, is pursuing new leads, but Garrett Phillip's murder remains unsolved.[75]

Commentary

Jewish Ethics

By Samuel J. Levine

Ethical questions revolving around the use – and potential misuse – of forms of inculpatory DNA evidence in criminal trials, though growing out of current scientific developments, represent but a recent manifestation of perennial philosophical questions that go to the heart of the nature of legal and moral judgment. From time immemorial, legal systems have grappled with the normative implications of new scientific and philosophical understandings of humanity and the world. Yet, the pace of technological progress has brought to the forefront a need to consider practical applications of issues that, until quite recently, seemed to remain within the purview of theoretical discussion, or perhaps science fiction.

As a system of thought that incorporates both law and theology, Jewish ethics examines interconnected conceptions of transcendental truth and practical reality. For example, going back millennia, Jewish legal philosophers have acknowledged the tension, if not the paradox, implicit in notions of free will and determinism. Positing God's omniscience, including foreknowledge of the future, amplifies questions of whether people should be judged based on actions that they have not yet committed. These puzzles, which are addressed in numerous sources of Jewish thought, are sometimes resolved through the axiomatic acceptance that God's judgment is inherently just, and therefore, Divine reward and punishment must take place within the realm of human exercise of free will.

It may not prove surprising that Jewish philosophers would premise an analysis of these issues on the axiom that God's laws are inherently just. Perhaps more surprisingly, though, many American judges and legal thinkers likewise willingly accept the doctrine of free will, as a matter of faith of sorts, rather than subjecting free will theories to just the kinds of rigorous debate applied to other complex areas of American law. As it turns out, American judges addressing issues of free will and determinism almost invariably invoke the striking recognition that their conclusions are based on principles and assumptions that may not – and need not – correlate with scientific advances and philosophical conceptions of truth.

Though perhaps in some ways unsatisfying, this approach to questions of free will may alternatively suggest a refreshing element of candour and humility on the part of the criminal justice system and the judges who mete out punishment. Judging others is a precarious, if unavoidable, pursuit, particularly in the context of criminal law, which ascribes moral culpability to those who are proven guilty. Although criminals may often be deserving of moral condemnation, a complete and accurate assessment of a person's moral worth remains outside the realm of human administration of justice, and beyond the grasp of limited human capabilities. Here too, Jewish thought has long acknowledged that notwithstanding the societal need to maintain order through the implementation of legal rules and consequences, ultimate moral judgment is reserved to the province of God.

The refusal among American judges to adopt philosophical or scientific approaches to determinism and free will may grow of an underlying adherence to the law's autonomy as representative of society's values and beliefs. To be sure, the law must take into account and should, as appropriate, employ advances in human understanding to inform and improve the workings of a legal system. Ideally, the law would progress alongside the emergence of scientific progress. Yet, the law is inextricably linked to society and reflective of human nature, which has all too often demonstrated an inability to harness scientific discoveries in a mode that promotes the underlying values of human progress. Among other lessons, prosecutorial abuse of DNA technology serves as a reminder of temptations and tendencies to exploit technology in the pursuit and exercise of power, in a manner that may overstep the bounds of ethics and justice. To progress alongside scientific advances, the law must incorporate a concomitant commitment to ongoing ethical advances as well.

Deontological Ethics

By Colin Marshall

The "Machine Models in Court" case raises a variety of moral questions from a deontological perspective. Two actions involving technology, in particular, call out for a deontological analysis: (1) District Attorney Fitzpatrick's seeking out a different result from that provided by TrueAllele (provided that was his intention) and (2) the analyst employing STRmix while referencing Hillary's profile as a known suspect in the case.

One traditional focus in deontological ethics has been on ruling out problematic forms of partiality. Problematically partial actions inappropriately advantage some people over others. Imagine someone who is contemplating whether to be a free rider, that is, they are considering whether to benefit from others' cooperation in some system while not cooperating themselves. Examples of free riding include using public transit without paying the fare and making use of government services while avoiding paying taxes. Such actions privilege the free rider's interest over others, and so (unless there are mitigating factors) show inappropriate partiality.

While no free riding occurs in the Machine Models in Court case, we can still ask whether Actions (1) and (2) would, as described, show problematic partiality. Unfortunately, there is no fully precise, uncontroversial way, of identifying where partiality exists or when it is problematic. However, many deontological ethicists have found it useful to use a procedure known as the universalization test.

The basic idea behind the universalization test is a familiar one, and is reflected in the question, "what if everyone did that?"; Slightly more precisely, the universalization test runs as follows: an agent morally evaluates a possible action by asking themselves whether they would endorse a system in which all agents in similar situations would act similarly. For example, a vendor who decides to lie in order to secure a lucrative contract might consider whether they would be willing to endorse a system in which *all* vendors would lie in order to secure lucrative contracts. In such a system, vendors would generally be known to be untrustworthy, which is something no vendor would want, so such lies would fail the universalization test. Though the universalization test appeals to an imaginary scenario, it helps reveal the vendor's actual partiality behind the lie, inappropriately privileging their own interests over others.

How would the universalization test apply to Action (1)? The question that DA Fitzpatrick should have asked himself would be something like this: would he endorse a system in which prosecuting attorneys always sought out another technological resource to support their antecedent view whenever the first resource did not? The answer to this question is less obvious than in the case of the lying vendor. However, if it is always (or almost always) possible to find some technological resource that supports *any* desired verdict, a similar problem would arise: in such a system, any given appeal to a resource would lose its persuasive value. No one who is attempting to appeal to a particular technological resource could endorse that general system. The unacceptability of this imaginary scenario would suggest that DA Fitzpatrick's action showed problematic partiality. On the other hand, if it is not always (or almost always) possible to find a technological resource that supports any desired verdict, such a system might not be problematic, suggesting that DA Fitzpatrick did not show problematic partiality.

How would the universalization test apply to Action (2)? Here, the question the analyst should have asked would be along these lines: would they endorse a system in which the application of algorithms in evaluating guilt was always (or almost always) informed by the analyst's antecedent beliefs and suspicions? Given how large role analysts have in the application of algorithms, this would threaten to make the appeal to technology like STRmix of little value, and so make it easy to support false accusations. Presumably, nobody could endorse a system in which any false accusation could easily be supported using technology. This would suggest that the analyst was showing unacceptable partiality in Action (2).

While the universalization test is useful in assessing partiality, it cannot be applied algorithmically. When the test is applied to a given action, the key question will always be about *which aspects of the action* should be generalized. For example, with Action (2), the question should not be whether the analyst would endorse a system in which everyone *who was suspicious of Nick Hillary* was allowed to use their own background beliefs in applying algorithms – that question would not help identify the relevant forms of partiality in this case. Hence, the test must always be guided by non-trivial judgments about which aspects of actions are morally relevant, and there is no simple formula for determining what those aspects are. Nonetheless, cases like the lying vendor suggest that these non-trivial judgments are sometimes relatively easy and uncontroversial. While most people are prone to certain forms of partiality, with some distance, many of us can recognize problematic partiality.

Notes

1. *Daubert* v. *Merrell Dow Pharmaceuticals, Inc.*, 509 U.S. 579 (1993), para. 597.
2. Liz Garbus, *Who Killed Garrett Phillips*, Part 1, HBO, 2019, 13:45.
3. Garbus, *Who Killed Garrett Phillips*, Part 1, 16:58.
4. Garbus, *Who Killed Garrett Phillips*, Part 1, 25:57.

5. United States Census Bureau, "Quick Facts: Potsdam Village, New York," U.S. Census Bureau Population Estimates, July 1, 2021 (v. 2021), https://www.census.gov/quickfacts/potsdamvillagenewyork, which states that, at present, the population of Potsdam is 83.2% Caucasian and 6.8% African American.

6. Garbus, *Who Killed Garrett Phillips*, Part 1, 31:30.

7. Garbus, *Who Killed Garrett Phillips*, Part 1, 44:44.

8. Garbus, *Who Killed Garrett Phillips*, Part 1, 45:25.

9. Garbus, *Who Killed Garrett Phillips*, Part 1, 37:49.

10. Garbus, *Who Killed Garrett Phillips*, Part 1, 9:30.

11. Garbus, *Who Killed Garrett Phillips*, Part 1, 9:30.

12. Garbus, *Who Killed Garrett Phillips*, Part 1, 9:30.

13. Garbus, *Who Killed Garett Phillips*, Part 2, 24:52.

14. Garbus, *Who Killed Garett Phillips*, Part 2, 33:26.

15. Garbus, *Who Killed Garett Phillips*, Part 2, 34:55.

16. Garbus, *Who Killed Garett Phillips*, Part 2, 55:10.

17. Garbus, *Who Killed Garrett Phillips*, Part 1, 12:24.

18. Bennett L. Gershman, "The Most Dangerous Prosecutor in New York," *Huffington Post*, September 30, 2016, updated September 20, 2017, https://www.huffpost.com/entry/the-most-dangerous-prosec_b_12085240.

19. Douglass Dowty, "Ex-North Country DA Who Mishandled Hillary Murder Trial Barred from Practicing Law," *Syracuse*, July 16, 2018, updated January 30, 2019, https://www.syracuse.com/crime/2018/07/ex-north_country_da_who_mishandled_hillary_case_barred_from_practicing_law.html.

20. Dowty, "Ex-North Country DA."

21. Gershman, "Most Dangerous Prosecutor."

22. *Brady v. Maryland*, 373 U.S. 83 (1963), requires prosecutors to disclose to the defendant any evidence bearing on guilt or innocence.

23. Gershman, "Most Dangerous Prosecutor."

24. Garbus, *Who Killed Garrett Phillips*, Part 2, 37:35.

25. Gershman, "Most Dangerous Prosecutor."

26. Jones stayed very close to Tandy throughout the process, even sitting beside her as she made statements to the other police officers.

27. Gershman, "Most Dangerous Prosecutor."

28. Gershman, "Most Dangerous Prosecutor."

29. Myrna S. Raeder, "See No Evil: Wrongful Convictions and the Prosecutorial Ethics of Offering Testimony of Jailhouse Informants and Dishonest Experts," *Fordham Law Review*, 76, no. 3 (2007): 1413–1452.

30. David Sommerstein, Lauren Rosenthal, and Brit Hansen, "Answers to Your Lingering Questions in the Nick Hillary Murder Trial," *North Country Public Radio*, September 26, 2016, https://www.northcountrypublicradio.org/news/story/32642/20160927/answers-to-your-lingering-questions-in-the-nick-hillary-murder-trial.

31. Sommerstein, "Answers to Lingering Questions." See also: New York Division of Criminal Justice Services, "Latent Search Service," *Division of Criminal Justice Services*, February 25, 2022, https://www.criminaljustice.ny.gov/ojis/safis.html.

32. Jesse McKinley. "Potsdam Boy's Murder Case May Hinge on Miniscule DNA Sample from Fingernail." *New York Times*, July 24, 2016, https://www.nytimes.com/2016/07/25/nyregion/potsdam-boys-murder-case-may-hinge-on-statistical-analysis.html.

33. Peter Gill, *Misleading DNA Evidence: Reasons for Miscarriages of Justice* (London: Elsevier, 2014), 3, 71.

34. Gill, *Misleading DNA Evidence*, 1.

35. Gill, *Misleading DNA Evidence*, 2.

36. Itiel E. Dror, and Greg Hampikian, "Subjectivity and Bias in Forensic DNA Mixture Interpretation," *Science & Justice*, 51, no. 4 (2011): 204–208.

37. M.D. Coble, J. Buckleton, J.M. Butler, T. Egeland, R. Fimmers, P. Gill, L. Gusmao, B. Guttman, M. Krawczak, N. Morling, et al., "DNA Commission of the International Society for Forensic Genetics: Recommendations on the Validation of Software Programs Performing Biostatistical Calculations for Forensic Genetics Applications," *Forensic Science International: Genetics*, 25 (2016): 191–197.

38. Scientific Working Group on DNA Analysis Methods (SWGDAM), "Guidelines for the Validation of Probabilistic Genotyping Systems," *Scientific Working Group on DNA Analysis Methods (SWGDAM) Guidelines* (March 2015), 2, https://www.swgdam.org/publications,

39. Scientific Working Group on DNA Analysis Methods (SWGDAM), "Guidelines," 2.

40. See e.g.: Susan A. Greenspoon, Lisa Schiermeier-Wood, and Brad C. Jenkins, "Establishing the Limits of TrueAllele Casework: A Validation Study," *Journal of Forensic Sciences*, 60, no. 5 (2015):1263–1276, doi: https://doi.org/10.1111/1556-4029.12810.

41. See *eg.*: Jo-Anne Bright, Duncan Taylor, Catherine McGovern, Stuart Cooper, Laure Russell, Damien Abnaro, and John Buckleton, "Developmental Validation of STRmix™, Expert Software for the Interpretation of Forensic DNA Profiles," *Forensic Science International: Genetics*, 23 (2016): 226–239, doi: https://doi.org/10.1016/j.fsigen.2016.05.007. Note also that validating probabilistic genotyping systems is regulated through guidelines put out by the Scientific Working Group on DNA Analysis Methods (SWGDAM). See: SWGDAM, SWGDAM Guidelines.

42. John Buckleton, "DNA Software in the Courts and the Future," *Jurist*, March 13, 2018, https://www.jurist.org/commentary/2018/03/john-buckleton-dna-software/.

43. James M. Curran, "A MCMC Method for Resolving Teo Person Mixtures," *Science & Justice*, 48, no. 4 (2008): 168–177, doi: https://doi.org/10.1016/j.scijus.2007.09.014.

44. Christophe Andrieu, Nando de Freitas, Arnaud Doucet, and Michael T. Jordan, "An Introduction to MCMC for Machine Learning," *Machine Learning*, 50 (2003): 5.

45. Andrieu, "Introduction to MCMC," 6.

46. Gill, *Misleading DNA Evidence*, 3.

47. Gill, *Misleading DNA Evidence*, 3.

48. See *eg.*: Melinda Matte, Linda Williams, Roger Frappier, and Johnathan Newman, "Prevalence and Persistence of Foreign DNA Beneath Fingernails," *Forensic Science International: Genetics*, 6, no. 2 (2012): 236–243, doi: https://doi.org/10.1016/j.fsigen.2011.05.008.

49. Gill, *Misleading DNA Evidence*, 3.

50. SWGDAM, "SWGDAM Guidelines."

51. SWGDAM, "SWGDAM Guidelines," 5.

52. SWGDAM, "SWGDAM Guidelines," 5, 8 *et seq.*

53. McKinley, "Potsdam Boy's Murder Case May Hinge on Miniscule DNA Sample."

54. *Frye* v. *United States*, 293 F. 1013 (D.C. Cir. 1923).

55. *Daubert* v. *Merrell Dow Pharmaceuticals, Inc.*, 509 U.S. 579 (1993). In Daubert, the court upheld Rule 702 of the Federal Rules of Evidence, which states that, "If scientific, technical, or other specialized knowledge will assist the trier of fact to understand the evidence or to determine a fact in issue, a witness qualified as an expert by knowledge, skill, experience, training, or education, may testify thereto in the form of an opinion or otherwise."

56. Garbus, *Who Killed Garett Phillips*, Part 2, 10:30.

57. Garbus, *Who Killed Garett Phillips*, Part 2, 11:30.

58. Cybergenetics, "STRmix Precluded as Inadmissible DNA Evidence in Hillary Case," *Cybergenetics*, August 26, 2016, https://www.cybgen.com/information/newsroom/2016/aug/STRmix-precluded-as-inadmissible-DNA-evidence-in-Hillary-case.shtml. Although many items were tested, it appears that no usable DNA was recovered from the majority of items.

59. Douglass Dowty, "DA Fitzpatrick Tests Cutting-Edge DNA in Potsdam Murder of 12-Year-Old Boy," *Syracuse*, July 25, 2016, last modified March 22, 2019, https://www.syracuse.com/crime/2016/07/da_fitzpatrick_tests_cutting-edge_dna_in_huge_potsdam_murder_trial.html.
60. Garbus, *Who Killed Garrett Phillips*, Part 1, 46:33.
61. Lyons, "Potsdam Murder Case."
62. See *eg.*: Jos Weusten, and Jos Herbergs, "A Stochastic Model of the Processes in PCR Based Amplification of STR DNA in Forensic Applications," *Forensic Science International: Genetics*, 6, no. 1 (2012): 17–25, doi: https://doi.org/10.1016/j.fsigen.2011.01.003.
63. Brendan J. Lyons, "Potsdam Murder Case Stretched Boundaries of DNA Rules," *Times Union*, September 10, 2016, https://www.timesunion.com/local/article/Potsdam-murder-case-stretched-boundaries-of-DNA-9215227.php.
64. Itiel E. Dror, William C. Thompson, Christian A. Meissner, Irv Kornfield, Dan Krane, Michael Saks, and Michael Risinger, "Context Management Toolbox: A Linear Sequential Unmasking (LSU) Approach for Minimizing Cognitive Bias in Forensic Decision Making," *Journal of Forensic Sciences*, 60, no. 4 (2015): 1111–1112, doi: https://doi.org/10.1111/1556-4029.12805.
65. Cybergenetics, "STRmix Precluded." Here, 'rfu' stands for 'relative fluorescence units'. The DNA alleles are measured using fluorescent tags, and the greater the fluorescence, the more of that size of DNA allele is present in the sample. An rfu that is below the cutoff is likely due to noise, and not an actual allele. The rfu cutoff should not be chosen arbitrarily, but should be selected after internal laboratory validation of similar samples run under similar conditions.
66. Cybergenetics, "STRmix Precluded."
67. *New York* v. *Oral Nicholas Hillary*, County Court, County of St. Lawrence, Decision & Order Re: DNA Analysis Admissibility, 2015-15, 3. In fact, no validation studies at all were performed by the New York State Crime Lab regarding STRmix, and it was not authorized to use STRmix.
68. *New York* v. *Oral Nicholas Hillary*, 4.
69. Cybergenetics, "STRmix Precluded."
70. *New York* v. *Oral Nicholas Hillary*, 4.
71. *New York* v. *Oral Nicholas Hillary*, 10.
72. Garbus, *Who Killed Garett Phillips*, Part 2, 1:36:15.
73. Garbus, *Who Killed Garett Phillips*, Part 2, 1:37:50.
74. Garbus, *Who Killed Garett Phillips*, Part 2, 1:37:50.
75. Garbus, *Who Killed Garett Phillips*, Part 2, 1:41:51.

5 Synthetic Media and Political Violence

Synthetic media therefore have the capacity to destroy social capital and credibility at its base, and this is equally true whether we accept their contents as being true or not. The more fundamental problem is that we do not have any objective criteria to determine what is worthy of belief. Moreover, drawing attention to the phenomenon of synthetic media only reinforces its effects, leaving us no choice but to fall back on our preferences, prejudices, and political ideologies[.]
~ (Tracey Dowdeswell and Sean Goltz)[1]

The Coup in Gabon

Gabon, a relatively stable country on the west coast of Africa, was rocked by an attempted coup in the early days of 2019. The coup itself was sparked, in large part, by a customary New Year's Eve address by President Ali Bongo Odimba that was posted on social media.[2] The President had been out of the public eye for some time – indeed he had not been in the country for over two months – sparking rumors that he was gravely ill or even dead.[3] It seemed to the opposition that he was not in charge of the country – that perhaps no one was in charge.[4] The government had not been honest and open about the health of the President.[5]

The New Year's message itself was widely denounced as a 'deepfake' – a colloquial term for artificial intelligence (AI)-manipulated media – in a political climate of widespread distrust of the ruling party, and an absence of publicly available and verifiable information about the President's condition.[6] While the New Year's message itself was unremarkable, the way that Ali Bongo is depicted in the video is decidedly odd: the President's face above his mouth is strangely immobile; he does not blink his eyes – which remain wide open throughout most of the address – for a full 1 minute and 39 seconds into the video; his speech and movements seem unnatural and stilted.[7] Comments on the video indicate that viewers felt it was "creepy" and a "deepfake."[8] On January 3, 2019, a prominent news outlet in Gabon published an article openly denouncing the video as a deepfake.[9]

'Synthetic media' is the technical term used to describe what are often called 'deepfakes'. Synthetic media make use of machine learning and neural

DOI: 10.1201/9781003293125-5

networks to create synthetic audio, photo, video, or even text media that appear authentic. Video footage can be altered to superimpose one person's speech and movements onto that of another.[10] Synthetic media go beyond photo and video footage and include AI-generated speech and text. These methods are being used to generate an increasing amount of the content we see on the internet and have been used by criminal actors to generate false e-mails, texts, or messages that look like they are from people who know you well.[11]

Synthetic text has even been used to generate academic articles that have been published in peer-reviewed journals. Several retractions in the Arabian Journal of Geosciences made Ivan Oransky's top 10 retractions of 2021 for a very unusual reason – they were pure gibberish.[12] One article had to be retracted from the Arabian Journal of Geosciences on the unusual grounds that "the content of this article is nonsensical."[13] The problem seems to have become quite widespread: more than 400 papers in journals owned by Springer Nature and hundreds more at Elsevier were also flagged.[14] Synthetic media technologies are developing faster than the technologies used to detect them, rendering deepfakes nearly invisible and highly manipulative.

The armed forces of Gabon – long in opposition to Bongo's ruling party – felt manipulated by the President's New Year's video, too. In the early morning hours of January 7, 2019, Lieutenant Kelly Ondo Obiang of the Republican Guard announced that the New Year's address showed that Ali Bongo was not fit to run the country.[15] Accordingly, the military was removing him from office and going to set up a national 'restoration council' to rule in his place.[16] Tanks rolled into the capital Libreville; internet service and power were cut.[17] About 300 protesters came out in support of the coup, and were assailed with teargas by government forces; gunshots rang out in the capital.[18] By 10:00 am, government forces were back in control, and the leaders of the coup were either dead or in custody.[19]

AI experts began to test whether the New Year's video had been manu-factured, and a clear consensus quickly emerged. Steve Grobman (the Chief Technology Officer at McAfee) ran the video through their algorithms and determined with a very high probability – at 92% – that the video was genu-ine.[20] Siwei Lyu, Professor of Computer Science at SUNY Albany, also exam-ined the video using his deepfake algorithm which scored even higher – at 99% – again supporting the authenticity of the video.[21] The algorithms did not find any evidence that *known* methods of synthesizing media had been used to produce the video.[22] If it was a deepfake, it must have been a very sophisticated one – a hypothesis that is not supported by the strangeness of the video itself.

The authenticity of the video was further confirmed by viewing Ali Bongo himself in subsequent public appearances, which show him to be very much alive – but also very much altered.[23] Alexander Dromeric, a neurologist con-sulted by the *Washington Post*, stated that Ali Bongo's movements and facial immobility were characteristic of someone who had had a stroke or suffered a brain injury of some kind.[24] The official government press office has since refused to confirm or deny whether Ali Bongo suffered a stroke.[25] The government of Gabon was indeed fueling misinformation and mistrust about the President's health – but by their lack of transparency and not by producing synthetic media. The problem with deepfakes is not merely that they generate false information, but that their very existence leads people to discredit reports which are, in fact, true.[26]

Aviv Ovadya, an expert in AI and synthetic media, states that deepfakes can be so dangerous precisely because they sow widespread mistrust of *all* media.[27] This gives rise to a kind of pernicious 'reality apathy' that Ovadya has termed the 'Infocalypse' – a fundamental loss of trust in social institutions.[28] An infocalypse arises at the point where we realize that we lack standards for truth and objectivity and that we have no control over fast-evolving technologies, as happened in Gabon. Ovadya states that "the stakes are high and the possible consequences more disastrous than foreign meddling in an election – an undermining or upending of core civilizational institutions, an 'infocalypse'."[29]

The reality apathy generated by synthetic media is costly – as Ovadya states, "either for media organizations forced to spend time and resources examining such videos or for societies that are thrown into debates about authenticity."[30] For Gabon, these costs were seen in political violence, widening fear, social fracture, and the deaths of two members of the rebel armed forces.

'Bulli Bai': Selling Women Synthetically in India

Elsewhere, we have described deepfakes as a particularly insidious form of computational propaganda,[31] mainly because of their potential to inflame tensions between nations, to endanger national security, and to undermine foreign policy and international diplomacy.[32] Recent events in India show how synthetic media can be used to deepen ethnic and religious animosities – in this case by directing politically-motivated sexual violence against Muslim women.

Rana Ayyub is a well-known investigative journalist in India. Not only is she an Indian woman negotiating public life in a socially conservative country, but she is a member of the Muslim minority – and one that has earned a reputation for criticizing members of the ruling Bharatiya Janata Party (BJP).[33] In April of 2018, she wrote a controversial report about the rape of an 8-year-old girl in Kashmir; she even went on the BBC and called out members of the nationalist BJP for marching in support of the accused.[34]

The blowback against Ayyub was swift but unpredictable. Unknown persons started circulating fake messages on Twitter purporting to come from Ayyub that expressed views such as, "'I hate India', 'I hate Indians', 'I love

Pakistan'," and "'I love child rapists and if they are doing it in the name of Islam I support them.'"[35] But worse was yet to come.

Someone from inside the BJP alerted Ayyub that there was a video being shared on WhatsApp that was going to be very difficult for her to see. "I'm going to send it to you but promise me you won't feel upset," they told her.[36] What Ayyub received was a pornographic video in which her face had been deepfaked onto the naked body of a – very young – woman.[37] She says that the video ended up on "almost every phone in India."[38] "You can call yourself a journalist, you can call yourself a feminist but at that moment, I just couldn't see through the humiliation," she said.[39] Even though no one thought (or was meant to think) that the faked pornography was real, it had its intended effect: "I've self-censored quite a bit out of necessity," Ayyub concluded.[40]

This is but one of many similar incidents in India. At the end of 2021, several women – all Muslim like Ayyub – appeared on a spoof auction site called 'Bulli Bai', which is a derogatory term for Muslim women. These women, too, were prominent in Indian public life: journalists, activists, and attorneys.[41] The site depicted them in derogatory, often sexually explicit or demeaning, circumstances. The images were sourced from social media accounts and then manipulated to depict the women in lewd positions.[42]

About 6 months earlier, a similar site called 'Sulli Deals' had gone viral.[43] As with Bulli Bai, these depictions would not meet the definition of a deepfake. They were 'shallowfakes' – synthetic media that isn't intended to be believed, but whose effects on their targets are keenly felt, nonetheless.

One of the women targeted on the fake auction sites is a 26-year-old student at Columbia University named Hiba Bég. Like Ayyub, Bég has been critical of the ruling party and its nationalist politics. She stated that this "intimidation is aimed at forcing Muslim women who raise their voices against the injustice to withdraw from public life …. But you don't back off, even if everything gets overwhelming."[44]

Ismat Ara, another victim of the Bulli Bai, tweeted out a screenshot of her being 'sold' at auction when she was listed as the 'Bulli Bai of the Day', remarking that her 2022 New Year had started with a "sense of fear & disgust."[45] In her case, too, the site used "a doctored picture of me in an improper, unacceptable and clearly lewd context."[46]

Another victim of the lewd auction sites is Quratulain Rehbar – a journalist from Indian-administered Kashmir and the wife of a sitting High Court Judge in Delhi.[47] Rehbar stated, "When I saw my photograph, my throat got heavy, I had goosebumps on my arms and I was numb. It was shocking and humiliating."[48] Rehbar stated that the fake auction site was intended "to degrade and humiliate vocal Muslim women."[49]

The fake auction sites also targeted Hasiba Amin, who works as a social media coordinator for the opposition Congress party.[50] She is concerned that the use of these sites to promote violence and threats against minority women will have consequences that go beyond their ability to humiliate and censor prominent Indian women.[51] She fears that online death threats and intimidation are going to fuel real-world sexual violence. She asks, "What guarantees do we have from the government that tomorrow the threats and intimidation online is not going to turn into the real-time sexual violence on the streets?"[52]

Sexual violence is a serious social and legal problem in India. Much of it is perpetrated against wives, due to the absence of marital rape laws, and adolescent girls.[53] There have also been incidents where rape, including gang rapes, have been ordered by village councils and used as tools of political violence.[54] Megha Kumar describes the anti-Muslim massacre in Gujarat in the summer of 2002, which saw the rape and mutilation of about 200 Muslim women by Hindu nationalist men; further examples abound.[55] This raises a serious question as to whether targeting politically active women in India through deepfake pornography might constitute a form of 'dangerous speech' as we have defined it in the case study concerning content moderation.

The Delhi police have recently charged two men in connection with the Bulli Bai and Sulli Deals auction sites, and their accounts were taken down by GitHub.[56] They were charged on March 4, 2022, with offences relating to sexual harassment and promoting enmity between ethnic and religious groups by the cybercrime Intelligence Fusion and Strategic Operations unit of the Delhi Police.[57] It remains to be seen what *real* impact this will have on the *virtual* sexual exploitation of women in India.

Actress Scarlett Johansson — one of the most-parodied women in the deep-fake porn industry — has stated that "trying to protect yourself from the internet and its depravity is basically a lost cause… a useless pursuit, legally, mostly because the internet is a vast wormhole of darkness that eats itself."[58]

Commentary

Utilitarian Ethics

By Peter Singer and Yip Fai Tse

The Key Consideration – The Value of Truth

We place a high value on truth and consider this value very difficult to override. Accordingly, we also value practices that embrace truth and intellectual honesty, such as the development of scientific methods that depend on evidence and on updating beliefs in the light of the best available evidence. In contrast, when we do not embrace truth, for example, if society observes a lot of misuse of deepfake technologies, we are likely to increase general mistrust in the authenticity of audio, videos, and images. This is shown in the case of the coup d'état in Gabon. And it will not only affect local politics. It is worrying that with deepfake technologies that are intended to deceive or mislead people and the ease of accessing them, there might be a point where anyone can be portrayed as doing anything, at any place (or in imaginary places). The consequences might include but are far from limited to, using any video or photo as evidence in court; both governments and NGOs not being able to identify violations of human rights;[59] and, flipping the thinking, those who commit crimes escaping conviction because they can plausibly deny the authenticity of the evidence against them.

Is It All About Deepfake?

To some, it might sound like deepfake is an entirely new problem, but it is not. Speaker of the House of Representatives of the U.S. Nancy Pelosi had been the target of a few altered videos to make her sound slurred or drunk, including her onstage speech at a Center for American Progress event. Those videos were rapidly spread and were used as evidence against her competence and work ethic.[60] It turned out that the videos were only slowed down to 0.75 speed, a function that is available in a wide variety of basic video playing or editing software.[61] Hence, we should not be tempted to think that deepfake technologies take all the responsibility. Nor should we think that deepfake is 'nothing more than Photoshop'. Using software like Photoshop takes quite some training and experience, maybe also talent, but deepfake technologies have been made into (too) easily assessable application programming interfaces (API), which makes it possible for millions of people to create deepfakes with a few clicks. Also, deepfake technologies are arguably more powerful than any previous methods of synthesizing media, as the current results have shown. And importantly, they will get still more sophisticated and harder to detect, as the algorithms behind them can continuously improve with more research and more training with data.

One might point to deepfake detection technologies as a reason why we should not worry too much. But they also have problems. First, they might not be 100% reliable. Second, even if they can detect deepfakes from real media with high confidence, it leaves detection out of the hands of people who cannot access the detection software. Third, it may lead us to reject valid evidence due to the widespread misuse of deepfakes, and the mistrust that this generated. Fourth, sometimes detection cannot undo the harm that is already done, such as in the case of unconsented pornography. Fifth, the detection methods might themselves help to improve deepfake technologies, either by forcing the designers or deepfake algorithms to become better, or even be directly employed to become tools for training these algorithms (e.g., as the discriminator of the Generative Adversarial Networks of training deepfake algorithms).

Some might raise the point that Utilitarianism would justify the use of deepfake even if it causes some harm to some individuals, as long as even more people would benefit from it. An example where people (seemingly not Utilitarians) have raised such an argument is unconsented synthetic pornography. We grant that the number of people who enjoy such materials far outnumbers the number of victims. But this argument is a travesty of what Utilitarians claim, though one perhaps generated by the misleading slogan "the greatest happiness of the greatest number." It is true that Jeremy Bentham, the founder of Utilitarianism, did use this slogan, but he later disavowed it, when he realized that it was being understood as suggesting that whatever benefits 51% of a population is right, even if the 51% benefit only slightly and the 49% suffer great harms. In the case discussed here, if deepfake pornography is made, the women who are portrayed by it will suffer harms that are of a totally different, and much more serious, kind than the loss of the "benefit" of not being able to view such pornography, if it was not made.

In addition to the long-term value of truth that we mentioned previously, there are also other important and long-term consequences from unconsented synthetic pornography: it reinforces and perpetuates the idea that women are objects who may be used, without their consent, for the enjoyment of others, and this will promote and make more acceptable attitudes that are harmful to women in many different aspects of their lives.

African Ethics

By John Murungi

It is difficult to think about the place of ethics in the world created by data science or, more specifically, by AI. I suspect that most data scientists and AI actors do not believe that they create a world in what they do or that they are residents of such a world. Many of them may be ignorant of this world and are its architects. It is also possible that some of them know that they are its architects and that they live in it. But even if they possess such knowledge, it is possible that they do not have a full grasp of its full ramifications.

It seems to me that what is required for meaningful determination of the place of ethics in the world created by data science and by AI is the deepening and broadening of the awareness of its ramifications. In the course of doing so, awareness of the existence of a point of view arising from a different world, a world that is pre-eminently ethical, can serve as an antidote to the dehumanizing aspects of the world generated by data science and by AI. It is in the context of such a point of view that I reflect on the link between synthetic media and political violence.

In the Gabon case the synthetically produced President Ali Bongo is difficult to distinguish from un-synthesized President Bongo. Similarly, the synthesized Indian woman journalist is difficult to distinguish from the un-synthesized Indian woman journalist. Even if there is a means to distinguishing one from the other, the means itself cannot remedy the harm that has already been done and there is no guarantee that further harm is not going to arise in the future. The difficulty of distinguishing what is real from what is fake will continue to haunt us.

Moreover, verification could be subject to AI manipulation. At times, the legitimacy of verification depends on what one wants to verify and what one wants to verify could be one's creation. One should not underestimate the persuasive power of synthetic media. AI has an untapped future. Its progress has no definite foreseeable future. It has ceaseless power to create its own future. In addition, it is possible that what is real as well as what is synthetic in the eyes of the beholder.

The world of synthetic media (the artificial world) could have the appearance of being more real than the real world. For some, it is increasingly becoming the real world. It should also be noted that ethics itself is not immunized from synthetic media manipulation. The good it aims at could be the good that is determined by AI. The harm that is perceived to be a product of AI may be perceived as its opposite: as the good. It is conceivable that those who produced a fake President in Gabon, as was the case with those who produced a fake Indian woman journalist, were not producing what was harmful. To this extent, there does not appear to be a necessarily causal link between AI and political violence.

The link between synthetic media and political violence calls attention to the link between data science and AI in politics. Neither the linkage nor what it entails is self-evident. The education of data scientists or the education of AI actors does not incorporate the study of politics. Superficially, it appears as apolitical, or as politically neutral. Even if the study of politics were to be included, most likely it would be political science in its positivistic sense. Most likely, consistent with other social sciences, it would be seemingly an apolitical political science.

Societies are politically established primarily for their well-being. This well-being is primarily the subject matter for ethics. If this is conceded, then the political cannot be divorced from the ethical and the ethical cannot be divorced from the political. The ethical is the political and the political is the ethical. When positivism determines the study of politics, the political science so determined has no place for ethics. Here, the political without a sense of the ethical has a home in the apolitical data science and in the apolitical in AI. Introducing ethics in data science and in AI education would appear as a distraction in this education. This education is often projected as value-free education – education without ethics, as apolitical education.

In Indigenous African societies, societal well-being is communal. This well-being is both political and ethical. It is well-being that has a place for individual well-being as well as group well-being. In either case, such well-being is not at the expense of society's well-being. This broader sense of societal well-being rests on the African concept of Ubuntu. In the Ubuntu mode of human being, it is claimed that "We are, therefore I am." It is recognized that in the pursuit of well-being, there is a place for an individual or for a group pursuit of well-being but not at the expense of the well-being of society. It is also recognized that an individual or a group could act contrary to societal well-being but there were societal controls to minimize the threat to society's well-being. These controls have weakened due to modern changes in society.

Changes in society from within and from without have complicated the construction of African societal well-being. The 'we' in Ubuntu has become complicated. One can no longer distinguish the real 'we' from the fake 'we'. Whatever was understood as the real 'we' has been subverted by the artificial 'we'. The detection of harm to society has become difficult because of the sophisticated role played by AI in manipulating perception. One of the AI consequences is the trespassing of societal borders. Today, no society in Africa or anywhere else is immunized from the ever-rising AI trespassing. It is increasingly and exceedingly difficult to distinguish what is real from what is fake or to distinguish what is harmless from what is harmful. Policing cyberspace is in order, but who supervises the police? Can AI actors serve as police officers? Who would supervise them? What is the political/ethical education they ought to have to enable them to do this in service in the pursuit of social justice? Has social justice itself not become an AI product? Are we not living in the age of AI tyranny?

Today, it appears that AI actors are the priests of our age. They function as political theologians. They are looked up to in society as the savior of humankind. AI centers have taken the aura of sanctity. Does that not call for new atheism? New infidels?

Indigenous Ethics

By Joey Miller and Andrea Sullivan-Clarke

The main moral question that we would like to address is posed at the end of this case study. Does "targeting politically active women in India through deepfake pornography … constitute 'dangerous speech'"? On an Indigenous understanding of ethics, absolutely.

To understand why, we should consider at least two sets of ethical issues that arise in answering this question. First, there is the issue of what is being used. In this case, the use of deepfake pornography raises ethical concerns over *how* and *why* such images were created and obtained. Second, there is the issue of what *results* from the use of deepfake pornography. In this case, the use of such images results in coercion, deception, and subjugation of politically active women in India. To explain why deepfakes constitute dangerous speech in this context, we will further explain both of these concerns in turn.

Regarding the ethical issue of how and why such images were created and obtained, the creation of such images constitutes a violation of sovereignty and autonomy. As is clear from the case of Synthetic Media and Political Violence, data can be weaponized. Even if synthetic, it can be used to harm, control, oppress, and disenfranchise others.

For Indigenous peoples, there is a history of data being weaponized.[62] As such, there is a movement by Indigenous researchers to gather and control data that pertains to their peoples.[63] The Data Sovereignty Movement is evidence of the importance of sovereignty and autonomy to Indigenous peoples. It is not just about how data can be weaponized, it is about who best knows what to do with the data. Indigenous peoples are in a better position to use (e.g., collect and implement) data about themselves to better their understandings and their lives than outsiders, who have often used such data for their own agendas.

The same can be said about the individuals in India whose likenesses are being used to create deepfake pornography. In this case, no consent was given to the use and creation of these instances of deepfake pornography. The data (i.e., the deepfakes) results from the use of someone's likeness without their consent. This is ethically wrong because a sufficient amount of sovereignty – the ability to control one's life and make their own decisions – is necessary to live well. Deepfakes result from a complete disregard of sovereignty.

In order to understand the issue of what results from the use of deepfake pornography, a key idea from Indigenous philosophy needs to be understood: *words have power*.[64] Given the connectedness of all things, the act of speaking, and the words spoken, have an effect on one's surroundings. This means not only do they have an effect on one's interactions, but they themselves *are* interactions (i.e., (inter)actions). For humans, words constitute a way of interacting with one's surroundings. Thus, they have an effect on the existence of harmony.

Clearly in the case of politically engaged women in India, their ability to live well is being affected by the use of deepfakes. The act of creating deepfakes, as well as the manner in which they are used, is a manipulative and coercive practice to get women to behave in a way that is "in line" or beneficial for those in power (e.g., men and the BJP). This is not done to live in harmony with one another – it is an attempt to force one group's will onto another while also promoting disharmony.

Notes

1. Tracey Dowdeswell, and Sean Goltz, "The Clash of Empires: Regulating Technological Threats to Civil Society," *Information & Communications Technology Law*, 29, no. 2 (2020): 194–217, doi: https://doi.org/10.1080/13600834.2020.1735060.
2. Gabon 24, *"Discours à la Nation du Président Ali Bongo Ondimba,"* Facebook (Video), December 31, 2018, https://www.facebook.com/watch/?v=324528215059254.
3. BBC News, "Gabon Coup Attempt: Government Says Situation Under Control," *BBC News*, January 7, 2019, https://www.bbc.com/news/world-africa-46779854.
4. Fact Checker, "The Suspicious Video that Helped Spark an Attempted Coup in Gabon," *Washington Post*, February 13, 2020, https://www.youtube.com/watch?v=F5vzKs4z1dc.
5. Fact Checker, "The Suspicious Video;" see also the original video: Gabon 24, *"Discours à la Nation."*
6. Fact Checker, "The Suspicious Video."
7. Fact Checker, "The Suspicious Video;" Gabon 24. *"Discours à la Nation."*
8. Gabon 24, *"Discours à la Nation."*
9. Fact Checker, "The Suspicious Video."

10. Simon Parkin, "The Rise of the Deepfake and the Threat to Democracy," *The Guardian*, June 22, 2019, https://www.theguardian.com/technology/ng-interactive/2019/jun/22/the-rise-of-the-deepfake-and-the-threat-to-democracy.

11. Aviv Ovadya, "Deepfake Myths: Common Misconceptions about Synthetic Media," *Alliance for Securing Democracy*, June 14, 2019, https://securingdemocracy.gmfus.org/deepfake-myths-common-misconceptions-about-synthetic-media/.

12. Retraction Watch, "The Top Retractions of 2021," *The Scientist*, December 21, 2021, https://www.the-scientist.com/news-opinion/the-top-retractions-of-2021-69533.

13. Editors, "Retraction Note: Neural Network-Based Urban Rainfall Trend Estimation and Adolescent Anxiety Management," *Arabian Journal of Geosciences,* 14, no. 2298 (2021), https://link.springer.com/article/10.1007/s12517-021-08732-6; retracting the original article: Retraction Note to: *Arabian Journal of Geosciences* (2021) 14: 1070. https://doi.org/10.1007/s12517-021-07246-5.

14. Retraction Watch, "The Top Retractions of 2021."

15. BNO News, "Soldiers in Gabon Try to Seize Power in Failed Coup Attempt," *BNO News*, January 7, 2019, https://bnonews.com/index.php/2019/01/military-coup-underway-in-gabon/.

16. Fact Checker, "The Suspicious Video."

17. BNO News, "Soldiers in Gabon."

18. BNO News, "Soldiers in Gabon."

19. BBC News, "Gabon Coup Attempt."

20. Fact Checker, "The Suspicious Video."

21. Fact Checker, "The Suspicious Video."

22. Fact Checker, "The Suspicious Video."

23. Fact Checker, "The Suspicious Video."

24. Fact Checker, "The Suspicious Video;" Gabon 24. "*Discours à la Nation.*"

25. Fact Checker, "The Suspicious Video."

26. Joplin Ty, "A Military Coup in Gabon Inspired by a Potential Deepfake Video is Our Political Future," *Al Bawaba News*, May 8, 2019, https://www.albawaba.com/news/military-coup-gabon-inspired-potential-deepfake-video-our-political-future-1284760.

27. Ali Breland, "The Bizarre and Terrifying Case of the 'Deepfake' Video that Helped Bring an African Nation to the Brink," *Mother Jones*, March 15, 2019, https://www.motherjones.com/politics/2019/03/deepfake-gabon-ali-bongo/.

28. Charlie Warzel, "He Predicted the 2016 Fake News Crisis. Now He's Worried about an Information Apocalypse," *Buzzfeed News*, February 11, 2018, https://www.buzzfeednews.com/article/charliewarzel/the-terrifying-future-of-fake-news#.auEWOgLKA.

29. Warzel, "He Predicted the 2016 Fake News Crisis."

30. Breland, "Bizarre and Terrifying."

31. Dowdeswell, "The Clash of Empires."

32. Danielle Keats Citron, "Prepared Testimony and Statement for the Record of Danielle Keats Citron, Morton and Sophia Macht Professor of Law, University of Maryland Carey School of Law, Hearing on "The National Security Challenge of Artificial Intelligence, Manipulated Media, and 'Deepfakes'" Before the House Permanent Select Committee on Intelligence, June 13, 2019, https://intelligence.house.gov/uploadedfiles/citron_testimony_for_house_committee_on_deep_fakes.pdf, 6.

33. Rana Ayyub, "I was the Victim of a Deepfake Porn Plot to Silence Me," *Huffington Post*, November 21, 2018, https://www.huffingtonpost.co.uk/entry/deepfake-porn_uk_5bf2c126e4b0f32bd58ba316.

34. Ayyub, "I was the Victim of a Deepfake."

35. Ayyub, "I was the Victim of a Deepfake."

36. Ayyub, "I was the Victim of a Deepfake."

37. Ayyub, "I was the Victim of a Deepfake."
38. Ayyub, "I was the Victim of a Deepfake."
39. Ayyub, "I was the Victim of a Deepfake."
40. Ayyub, "I was the Victim of a Deepfake."
41. Samir Yasir, "Online 'Auction' Is Latest Attack on Muslim Women in India," *New York Times*. January 3, 2022, https://www.nytimes.com/2022/01/03/world/asia/india-auction-muslim-women.html.
42. RT, "Women 'Auctioned' Online in India Through 'Lewd' Photos – Reports," *RT*, January 2, 2022, https://www.rt.com/news/545003-women-auctioned-online-india/.
43. Yasir, "Online Auction."
44. Yasir, "Online Auction."
45. RT, "Women 'Auctioned' Online."
46. RT, "Women 'Auctioned' Online."
47. Al Jazeera Online, "India's Muslim Women Again Listed on App for 'Auction'," *The Daily Star*, January 4, 2022, https://www.thedailystar.net/news/asia/india/news/indias-muslim-women-again-listed-app-auction-2931926.
48. Al Jazeera Online, "India's Muslim Women."
49. Al Jazeera Online, "India's Muslim Women."
50. Yasir, "Online Auction."
51. Yasir, "Online Auction."
52. Yasir, "Online Auction."
53. Anita Raj and Lotus McDougal, "Sexual Violence and Rape in India," *Lancet*, 383, no. 9920 (2014): 865, doi: https://doi.org/0.1016/S0140-6736(14)60435-9.
54. Raj, "Sexual Violence," 865.
55. Megha Kumar, *Communalism and Sexual Violence in India: The Politics of Gender, Ethnicity and Conflict* (London: IB Taurus & Co., 2016), 1.
56. NewsWire, "Delhi Police File Chargesheets in 'Bulli Bai', 'Sulli Deals' Cases," *CanIndia*, March 8, 2022, https://www.canindia.com/delhi-police-file-chargesheets-in-bulli-bai-sulli-deals-cases/.
57. NewsWire, "Delhi Police File Chargesheets."
58. D. Harwell, "Scarlett Johansson on Fake AI-Generated Sex Videos: 'Nothing Can Stop Someone from Cutting and Pasting My Image'," *Washington Post*, December 31, 2018, https://www.washingtonpost.com/technology/2018/12/31/scarlett-johansson-fake-ai-generated-sex-videos-nothing-can-stop-someone-cutting-pasting-my-image/.
59. It is not that we think animal right issues shouldn't be scrutinized in the same way. It is rather that, animal abuses don't enter the stage of international politics.
60. Drew Harwell, "Faked Pelosi Videos, Slowed to Make Her Appear Drink, Spread Across Social Media," *Washington Post*, May 24, 2019, https://www.washington-post.com/technology/2019/05/23/faked-pelosi-videos-slowed-make-her-appear-drunk-spread-across-social-media/.
61. Drew Harwell, "Faked Pelosi Videos, Slowed to Make Her Appear Drink, Spread Across Social Media," *Washington Post*, May 24, 2019, https://www.washington-post.com/technology/2019/05/23/faked-pelosi-videos-slowed-make-her-appear-drunk-spread-across-social-media/.
62. e.g., blood quantums, boarding and residential schools, treaties that were fraudulently signed, etc.
63. See e.g.: Tahu Kukutai, and John Taylor, *Indigenous Data Sovereignty: Toward an Agenda* (Canberra: Australian National University Press, 2016).
64. Doug Good Feather, *Think Indigenous: Native American Spirituality for a Modern World* (New York City: Hay House, Inc., 2021), 38.

6 Biometrics and Facial Recognition

> [I]ndividuals do not forego their right to privacy merely by participating in the world in ways that may reveal their face to others, or that may enable their image to be captured on camera. Privacy is vital to dignity, autonomy, personal growth and the free and open participation of individuals in democratic life. When surveillance increases, individuals can be deterred from exercising these rights and freedoms.[1]
>
> ~ (Daniel Therrien, Privacy Commissioner of Canada)

Clearview AI

Online child sexual exploitation is a serious and growing problem around the world. Between 2014 and 2019, reports to the Royal Canadian Mounted Police (RCMP) of photos or videos depicting child sexual abuse increased by about 1106%.[2] In 2019, the RCMP received a whopping 102,967 reports of online child sexual abuse.[3] Given the scale of the problem, it can be difficult to locate, identify, and take down the material.

This can be particularly tough on the victims. Pornographic materials depicting their abuse can linger on the internet for years. In Chapter 5 we discussed how fake pornographic videos have been traumatizing for female journalists in India, and that the posting of such materials has been used as a tool of political violence against women in that country. For the victims of child sexual abuse, the videos depicting them in degrading and abusive sexual circumstances are *real*, and they don't seem to ever go away.

Several non-profits have begun to apply machine learning technologies to the problem of identifying and taking down material depicting child sexual abuse. In the United States, the non-profit Thorn has developed a tool called Spotlight that uses facial recognition technology (FRT) to identify victims of child sexual exploitation and trafficking.[4] Law enforcement officers can upload a photo of a missing or exploited child, and then search for videos depicting the child, or online ads offering the child for sex.[5] In Canada, Project Arachnid uses artificial intelligence (AI) to search for images of child sexual exploitation on the clear and dark webs and then issues takedown requests to have the material removed.[6] They have removed about 6 million photos and videos from the web, and more are posted every day.[7]

DOI: 10.1201/9781003293125-6

In October of 2019, the National Child Exploitation Crime Centre (NCECC) – a division of the RCMP – began to use facial recognition technology to identify child victims of online sexual exploitation.[8] They purchased two licenses from a U.S. company named Clearview AI which would give them access the Clearview's facial recognition algorithms and massive database of photos.[9] They also used several free trial accounts that were offered by Clearview to law enforcement agencies.[10] The RCMP used the technology at the National Headquarters, as well as in British Columbia, Alberta, Manitoba, and New Brunswick.[11]

The NCECC states that it used Clearview's facial recognition technology in 15 cases and rescued 2 children.[12] Further, Clearview was used about 14 times to identify a perpetrator evading law enforcement.[13] The RCMP states that it otherwise used the technology on a trial basis to see what utility facial recognition technology might have in advancing criminal investigations more generally.[14] The Privacy Commissioner states that the RCMP did not disclose the purpose of most of the hundreds of searches that were performed.[15]

The Toronto Police also used Clearview in several investigations around the same time. As with the RCMP, the use was not authorized or subject to any internal controls.[16] In the end, it was deployed in 84 investigations between October 2019 and February 2020, most of which involved homicides and sex crimes: they identified 12 victims – 10 of whom were children[17] – as well as 2 witnesses, and 4 suspects.[18] The Toronto Police Services Board states that it has no plan to use Clearview again,[19] and has recently instituted a policy governing the use of facial recognition technology.[20]

Like the Toronto Police Service,[21] the RCMP initially denied using Clearview to Canada's Privacy Commissioner.[22] The Office of the Privacy Commissioner then launched an investigation into whether the use of Clearview violated Canada's privacy laws. Several concerns were raised by the Privacy Commissioner, including the fact that facial recognition technology might be deployed against activists and protesters,[23] that it has the potential to be "a highly invasive surveillance technology,"[24] and that its use might violate other fundamental human rights, including by fostering racial discrimination in the criminal justice system.[25]

Biometrics and Facial Recognition Technology

Facial recognition is one form of biometric technology – many of which are now facilitated by machine learning. Biometric identifications are more secure than other forms of ID, like passwords, cards, and other tokens – all of which can be lost, stolen, spoofed, or subject to trial and error attacks (themselves made easier by AI systems deployed by bad actors).[26] A reliable biometric will be both highly individualizing – it must accurately distinguish between one person and another – and stable over time, so that an individual can rely on being able to use their identification when needed.[27] Two of the most reliable and individualizing biometrics are fingerprints and iris scans, and this is likely why they were chosen for India's Aadhaar program.[28] However, reliable does not mean foolproof. Iris scans can be difficult for the elderly and others with cataracts;[29] fingerprints are an unreliable form of ID for about 1–3% of the population[30] and, as many Indians have discovered to their detriment, they must be updated as our fingerprints change over our life spans, while for others can be altered or lost entirely due to hard manual labor.[31] At the same time, what makes biometrics one of the most desirable and secure forms of identification – the fact that they are an integral part of who we are as unique individuals – is precisely what makes the information they contain so sensitive.[32]

Facial recognition technology is a less reliable form of identification than fingerprints and iris scans. Facial recognition technologies do not measure or analyze our faces directly. They "have no built-in notion of a particular person" and "are not built to identify particular people."[33] Instead, they measure certain distances between facial features: the distance between our eyes, the width of our nose, the depth of our eye sockets, the length of our jawline, etc.[34] These measurements produce a vector value that serves as a proxy for a given individual.[35]

A comparison must then be made between the vector values of different images that are being compared. Comparisons can be made for enrolment purposes, for example, when taking a photograph to enroll in a facial recognition program that would give the person access to a secure location, an electronic device, or receive a service such as a prescription medication. Comparison can also be on a 'one-to-one' basis, such as when a person uses an image of their face to later gain access to the program.[36] Searches can also be 'one-to-many,' which involve comparing the person's photo against a database of similar photos.[37] The vector value created from the individual's image would then be compared to the image in the database to determine the probability that the images are of the same person. A provisional identification is made when the probability exceeds a certain threshold value.[38] This is checked by a human operator who compares the images, but this can introduce a form of confirmation bias into the identification process. The human operator is more likely to select one of the candidates and to choose the highest-ranked candidate because they believe the algorithm is highly effective.[39] Police use of images to search for suspects, and the RCMP's use of Clearview AI to search for victims of online sex crimes would involve one-to-many searches of this type.

Facial Recognition and Privacy

The Privacy Commissioner found that Clearview AI had violated Canada's privacy laws when they collected Canadians' private information without their consent by putting together an image database of about 10 billion photographs for comparison.[40] Clearview countered that these photos, which had been scraped from the internet and social media sites, had been made publicly available when they were posted. The RCMP, in turn, relied on Clearview's assertions that the photos had been made 'public'.[41] The Privacy Commissioner found that the RCMP would have had to have obtained the subjects' consent to use their images in their investigations.[42]

In the end, the RCMP agreed to change its policies and abide by the Privacy Commissioner's recommendations.[43] By that time, the software had been tested in Vancouver, Edmonton, Calgary, and Ottawa.[44] It has also been used in France, the United States, Australia, and the UK, where it has been the subject of much controversy and several legal actions over its potential for mass surveillance and unreasonable searches and seizures.[45] The fact that the data used to train and operate a biometric algorithm has been made public by users does not entitle data scientists to assume that the owners of that data have consented to its use.

Commentary

Indigenous Ethics

By Joey Miller and Andrea Sullivan-Clarke

Biometrics and facial recognition technology have the potential to be tools of oppression. This is especially the case when there is no oversight or policy regarding their use. Without such guidance, the technology may be used arbitrarily against marginalized communities, particularly those that challenge the status quo. In the absence of external oversight and the creation of policies regarding the use of biometrics, a primary concern for Indigenous individuals and communities is the possibility of being targeted for surveillance. This is a legitimate worry – it is not the first time that militarized responses have been utilized against Indigenous people. The recent histories of the United States and Canada are replete with examples of militarized responses to Indigenous protests, such as the Oka Crisis, Standing Rock, and the Wet'suwet'en Blockade.[46] Colonialism can be described as war and the methods used by federal governments to preserve the status quo reflect this.[47] However, the worry concerning surveillance and the loss of one's privacy in the Case Study is a Western framing of the issue.[48]

To understand the ethics of using biometrics and facial recognition technology in a way that is consistent with an Indigenous worldview requires reframing the discussion in terms of the relations that are involved instead of as a question of individual rights. For example, the use of biometrics and facial recognition technology impacts *the relationship* between Indigenous people and the federal government. When governments emphasize truth and reconciliation while their agencies employ technology to surveil Indigenous individuals and communities, it negatively impacts a historically tense relationship. When the actions and words of either party are not consistent, both sides in the relationship suffer: the work for decolonization is made all the harder for Indigenous people, and governments (like the United States and Canada) not only lose face on the world's stage and to the Indigenous peoples, but they also fail to benefit from the diverse approaches to problem-solving that an Indigenous worldview may provide.

Biometrics and facial recognition technology, from an Indigenous point of view, are not immoral in themselves. Rather, the way the technology is used (e.g., does its use contribute to balance and harmony? Or does its use promote good relations?) determines how one relates to it. The lack of oversight and policies concerning how the technology should be used allows for the arbitrary use against marginalized communities. Not only should a policy of use be created, but it should be done in consultation with the Indigenous people and communities.[49] From the Western perspective, Clearview AI is a tool and its application is limited to the use of governmental policing agencies. Consultation with Indigenous communities may yield other uses, such as restoring balance by addressing the numerous cases of missing and murdered Indigenous women and girls (MMIWG). Working in collaboration with Indigenous people contributes to the relationship of trust while preserving the sovereignty and self-determination of Indigenous nations.

Another issue that requires consideration is that some biometric measures, like fingerprints, and facial recognition technology, may inspire a false confidence in the identification of individuals believed to have committed crimes. Overconfidence with respect to the use of this technology indicates a lack of humility (an Indigenous value). Having such confidence will negatively impact Indigenous people as they are already overrepresented in the incarceration rates of both the United States and Canada.[50] Searches that rely on the comparison of a photo against those in a database call into doubt the objectivity of the individual making the comparisons. Individuals with phenotypical features associated with racial groups will be made more vulnerable to the prejudices and implicit biases of those making the determinations. Humility may be employed to mitigate the issue of overconfidence.

If biometrics and facial recognition technology are potential tools of oppression, then there should be oversight and/or the creation of policies regarding their use. A policy developed solely from a Western worldview will lack the resources to mitigate oppression because it considers the technology a tool and not worthy of standing in relation to it. Collaborating with local Indigenous communities and applying Indigenous values by asking how the technology contributes to harmony or whether it introduces chaos, would not only strengthen relations between nations, but it would also provide a more robust understanding of the technology. Thus, federal governments and their agencies should rethink how they envision using these tools. They should collaborate with local Indigenous communities to obtain a robust understanding of this technology and develop policies that prevent future abuses of marginalized communities.

Humanitarianism and the Laws of Armed Conflict

By Tracey Dowdeswell

Despite its legal challenges, Clearview AI is actively seeking its first big U.S. government contracts – particularly with federal law enforcement agencies such as the FBI, Immigration & Customs Enforcement, and the Fish & Wildlife Service.[51] They are also researching the use of facial recognition and augmented reality to secure Air Force base checkpoints.[52]

Despite being untested in armed conflict, Clearview AI officially rolled out its technology for use in wartime on March 10, 2022, in Ukraine. Ukraine's Defence Ministry began using Clearview AI, ostensibly to identify the dead, to combat misinformation, and to "vet people of interest at checkpoints."[53] Given the proneness of facial recognition systems to error and bias, its use against civilians is particularly disturbing, since false positives may lead to wrongful detentions or even killings in violation of the international laws of war. If the system misidentifies civilians at checkpoints, or in and around battle sites, then this could result in the commission of war crimes.[54]

Distinguishing civilians from combatants in wartime – including civilians who may act as resistance fighters, spies, rebels, or other guerrilla forces – and how this should be dealt with in the international laws of war is a problem of long-standing.[55] The current state of the law is that military forces may only target such civilians as long as they actively pose an armed threat; any other activities must be dealt with by a fair trial under domestic laws for those persons who are not entitled to be treated as prisoners of war.[56] For the past 150 years, states have pushed back against international humanitarian instruments, such as *The Hague* and *Geneva Conventions* and their *Protocols Additional*, to give them more latitude to identify and kill civilians whom they think may pose a threat to their war effort.[57]

The use of facial recognition technology in a war zone certainly has the potential to exacerbate this long-standing problem and to facilitate war crimes. A civilian approaching a checkpoint, or seeking humanitarian assistance, may be targeted by the armed forces because of a false positive match from a facial recognition database. When this technology was offered to the Ukrainian military, no efforts were put in place to ensure that it would not be used to commit war crimes of this type.[58] Indeed, Clearview AI themselves noted that the "exact purpose for which Ukraine's defense ministry is using the technology is unclear."[59] Clearview's CEO, Hoan Ton-That, stated that he would never want to see the technology used in violation of the *Geneva Conventions* and that it should never be used as the sole source of identification, but he has not put in place any rules or safeguards that would prevent this from happening.[60]

The international laws of war would generally prohibit the "acquisition or adoption" of novel methods and means of warfare unless the State can first determine "whether its employment would, in some or all circumstances, be prohibited by this Protocol or by any other rule of international law[.]"[61] The use of facial recognition technology in an international armed conflict is novel, and given the very high risks involved in using facial recognition technologies to identify unknown persons in real-time, along with their proven potential for errors and demographic bias, we argue that it should generally be considered a prohibited means of warfare. We recommend that international law be interpreted such that biometric identifiers like facial recognition technologies can only be deployed in international armed conflicts for humanitarian purposes, such as identifying decedents and displaced persons and reuniting them with their families.

Notes

1. Officer of the Privacy Commissioner of Canada, "Police Use of Facial Recognition Technology in Canada and the Way Forward." Special Report to Parliament in the OPC's Investigation into the RCMP's Use of Clearview AI and Draft Joint Guidance for Law Enforcement Agencies Considering the Use of Facial Recognition Technology," June 10, 2021, https://www.priv.gc.ca/en/opc-actions-and-decisions/ar_index/202021/sr_rcmp/#toc1.
2. Royal Canadian Mounted Police, "RCMP Use of Facial Recognition Technology," *Royal Canadian Mounted Police (Press Release)*, February 27, 2020, https://www.rcmp-grc.gc.ca/en/news/2020/rcmp-use-facial-recognition-technology.
3. Royal Canadian Mounted Police, "RCMP Use of Facial Recognition Technology."
4. Tom Simonite, "How Facial Recognition is Fighting Child Sex Trafficking," *Wired*, June 19, 2019, https://www.wired.com/story/how-facial-recognition-fighting-child-sex-trafficking/.
5. Simonite, "How Facial Recognition Is Fighting Child Sex Trafficking."
6. Canadian Centre for Child Protection, "Programs & Initiative: Project Arachnid," *Canadian Centre for Child Protection*, 2022, https://protectchildren.ca/en/programs-and-initiatives/project-arachnid/.
7. Canadian Centre for Child Protection, "Programs & Initiative: Project Arachnid."
8. Office of the Privacy Commissioner of Canada, "Report of Findings: Investigation into the RCMP's Collection of Personal Information from Clearview AI (Involving Facial Recognition Technology)," Complaint under the Privacy Act, *Office of the Privacy Commissioner of Canada*, June 10, 2021, https://www.priv.gc.ca/en/opc-actions-and-decisions/ar_index/202021/sr_rcmp/#toc1, para. 15.
9. Office of the Privacy Commissioner of Canada, "Report of Findings," para. 1.
10. Office of the Privacy Commissioner of Canada, "Report of Findings," para. 1.
11. Office of the Privacy Commissioner of Canada, "Report of Findings," para. 14.
12. Royal Canadian Mounted Police, "RCMP Use of Facial Recognition Technology."
13. Office of the Privacy Commissioner of Canada, "Report of Findings," para. 16.
14. Royal Canadian Mounted Police, "RCMP Use of Facial Recognition Technology."
15. Office of the Privacy Commissioner of Canada, "Report of Findings," para. 5.
16. Nicole Brockbank, "Toronto Police Used Clearview AI Facial Recognition Software in 84 Investigations," *CBC News*, December 23, 2021, https://www.cbc.ca/news/canada/toronto/toronto-police-report-clearview-ai-1.6295295.
17. Wendy Gillis, "144 Toronto Police Officers Signed Up to Use Clearview AI 'Mass Surveillance' Tech," *Toronto Star*, February 28, 2022, last modified March 1, 2022, https://www.thestar.com/news/gta/2022/02/28/144-toronto-police-officers-signed-up-to-use-clearview-ai-mass-surveillance-tech.html.
18. Brockbank, "Toronto Police Used Clearview."
19. Brockbank, "Toronto Police Used Clearview."
20. Gillis, "144 Toronto Police Officers."
21. Kate Schneider, "Use of Controversial Surveillance Technology Demonstrates the Need to Limit Power," *CBC News*, January 8, 2022, https://www.cbc.ca/news/opinion/opinion-police-facial-recognition-technology-clearview-ai-1.6306357.

22. Office of the Privacy Commissioner of Canada, "Report of Findings," para. 10.
23. Office of the Privacy Commissioner of Canada, "Report of Findings," para. 9.
24. Office of the Privacy Commissioner of Canada, "Draft Privacy Guidance on Facial Recognition for Police Agencies," *Office of the Privacy Commissioner of Canada*, n.d., accessed March 25, 2022, https://www.priv.gc.ca/en/about-the-opc/what-we-do/consultations/gd_frt_202106/, para. 5.
25. Office of the Privacy Commissioner of Canada, "Draft Privacy Guidance, para. 11.
26. John D. Woodward Jr., Nicholas M. Orlans, and Peter T. Higgins, *Biometrics: Identity Assurance in the Information Age* (New York: McGraw Hill, 2003), 12.
27. Woodward, *Biometrics*, 28.
28. Khera Reetika, "Aadhaar Failures: A tragedy of Errors," *Economic & Political Weekly*, 54, no. 14 (2019), https://www.epw.in/engage/article/aadhaar-failures-food-services-welfare; Woodward, *Biometrics*, 89.
29. Anuradha Varanasi, "Elderly at a Disadvantage as Finger Prints Not Recognizable for Aadhaar," *The Week*, August 2, 2017, https://www.theweek.in/content/archival/news/india/elderly-disadvantage-finger-prints-not-recognisable-aadhaar.html.
30. Woodward, *Biometrics*, 64.
31. Varanasi, "Elderly at a Disadvantage; see also: Ramnath, *The Aadhaar Effect*, Chapter 3: Loc. 241/361.
32. Woodward, *Biometrics*, xxv.
33. Patrick Grother, Mei Ngan, Kayee Hanaoka, "Face Recognition Vendor Test (FRVT) Part 3: Demographic Effects," National Institute of Standards and Technology Interagency or Internal Report, NISTIR 8280, December 2019, https://doi.org/10.6028/NIST.IR.8280, 4.
34. Office of the Privacy Commissioner of Canada, "Draft Privacy Guidance," para. 26.
35. Grother, "Face Recognition Vendor Test," 4.
36. Grother, "Face Recognition Vendor Test," 4. One-to-one matching can also be used to exclude a person from a location or service, such as when a person enrolls their photograph in a program to deny them admittance to a casino or other gambling facility.
37. Grother, "Face Recognition Vendor Test," 5.
38. Grother, "Face Recognition Vendor Test," 4.
39. Grother, "Face Recognition Vendor Test," 69.
40. Kate Schneider, "Use of Controversial Surveillance Technology."
41. Office of the Privacy Commissioner of Canada, "Report of Findings: Investigation into the RCMP's Collection of Personal Information from Clearview AI (Involving Facial Recognition Technology)," Complaint under the Privacy Act. *Office of the Privacy Commissioner of Canada*, June 10, 2021, https://www.priv.gc.ca/en/opc-actions-and-decisions/ar_index/202021/sr_rcmp/#toc1, para. 40.
42. Office of the Privacy Commissioner of Canada, "Report of Findings," para. 3.
43. Office of the Privacy Commissioner of Canada, "RCMP's Use of Clearview AI's Facial Recognition Technology Violated *Privacy Act*, Investigations Concludes," *Office of the Privacy Commissioner of Canada* (Press Release), June 10, 2021, https://www.priv.gc.ca/en/opc-news/news-and-announcements/2021/nr-c_210610/.
44. Schneider, "Use of Controversial Surveillance Technology."
45. Schneider, "Use of Controversial Surveillance Technology."
46. See: Melinda Meng, "Bloody Blockades: The Legacy of the Oka Crisis," *Harvard International Review*, June 30, 2020, https://hir.harvard.edu/bloody-blockades-the-legacy-of-the-oka-crisis/; Jamil Dakwar, "Why Did a Private Security Contractor Treat Standing Rock Protesters like 'Jihadists'?" *American Civil Liberties Union*, June 2, 2017, https://www.aclu.org/blog/free-speech/rights-protesters/why-did-private-security-contractor-treat-standing-rock; Jessica McDiarmid, "Heavily Armed Mounties Dismantle Wet'suwet'en Blockade," *National Observer*, November 19, 2021, https://www.nationalobserver.com/2021/11/19/news/heavily-armed-mounties-dismantle-wetsuweten-blockade.

47. See: Scott Demuth, "Colonization is Always War," in *For Indigenous Minds Only: A Decolonization Handbook.* ed. Waziyatawin and Michael Yellow Bird (Santa Fe, New Mexico: School for Advanced Research Press, 2012).
48. By pointing this out, I am not asserting that Indigenous people do or do not value privacy. Rather, I contend that individual privacy is an artifact of a western liberal democratic society.
49. Arguably, consultation would be best for all social groups that may be oppressed by the arbitrary use of this technology. However, given that Indigenous communities in the United States and Canada are sovereign nations, and there are policies of consultation already in place, I suggest consultation between Indigenous nations and federal governments as a minimum for developing plans for oversight and policy for the use of bioinformatics and facial recognition technology.
50. See: Scott Clark, *Overrepresentation of Indigenous People in the Canadian Criminal Justice System: Causes and Responses* (Ottawa: Department of Justice Canada, 2019), accessed July 1, 2022, https://www.justice.gc.ca/eng/rp-pr/jr/oip-cjs/oip-cjs-en.pdf; Leah Wang, "The U.S. Criminal Justice System Disproportionately Hurts Native People: The Data, Visualized," *Prison Policy,* October 8, 2021, https://www.prisonpolicy.org/blog/2021/10/08/indigenouspeoplesday/.
51. Paresh Dave, and Jeffrey Dastin, "Exclusive: Facial Recognition Company Clearview AI Seeks First Big Deals, Discloses Research Chief," *Reuters,* February 22, 2022, https://www.reuters.com/technology/exclusive-facial-recognition-company-clearview-ai-seeks-first-big-deals-2022-02-22/.
52. Dave "Exclusive: Facial Recognition Company Clearview AI."
53. Paresh Dave, and Jeffrey Dastin, "Exclusive: Ukraine has Started Using Clearview AI's Facial Recognition During War," *Reuters,* March 14, 2022, https://www.reuters.com/technology/exclusive-ukraine-has-started-using-clearview-ais-facial-recognition-during-war-2022-03-13/.
54. Dave, "Exclusive: Ukraine has Started Using Clearview."
55. Tracey Leigh Dowdeswell, "The Brussels Peace Conference of 1874 and the Modern Laws of Belligerent Qualification," *Osgoode Hall Law Journal,* 54, no. 3 (2017): 806.
56. International Committee of the Red Cross (ICRC), *Protocol Additional to the Geneva Conventions of 12 August 1949, and relating to the Protection of Victims of Non-International Armed Conflicts (Protocol II),* 8 June 1977, 1125 UNTS 609, https://www.refworld.org/docid/3ae6b37f40.html, Article 4(2)(a) states that violence to civilians and those who have ceased to take part in hostilities is prohibited; see also: International Committee of the Red Cross (ICRC), *Protocol Additional to the Geneva Conventions of 12 August 1949, and relating to the Protection of Victims of International Armed Conflicts (Protocol I),* 8 June 1977, 1125 UNTS 3, https://www.refworld.org/docid/3ae6b36b4.html, Article 45(2); see also Articles 46 and 47 of Protocol I dealing with those suspected of being spies or mercenaries, in which they may be denied prisoner of war status if they are not regular combatants, and should be tried under domestic laws.
57. Dowdeswell, "The Brussels Peace Conference."
58. United Nations General Assembly, *Rome Statute of the International Criminal Court (last amended 2010),* 17 July 1998, ISBN No. 92-9227-227-6, https://www.refworld.org/docid/3ae6b3a84.html; not only are such acts prohibited under the *Geneva Conventions* and their *Protocols Additional, supra* note 82, but are identified as war crimes in the *Rome Statute,* particularly in in Article 8(2)(a)(i) and Article 8(2)(b)(i).
59. Dave, "Exclusive: Ukraine has Started Using Clearview."
60. Dave, "Exclusive: Ukraine has Started Using Clearview."
61. International Committee of the Red Cross (ICRC), *Protocol Additional to the Geneva Conventions of 12 August 1949, and relating to the Protection of Victims of International Armed Conflicts (Protocol I),* Article 36.

7 Content Moderation

Hate Speech and Genocide in Myanmar

... Here is a racist joke. Here is a man having sex with a farm animal. Here is a graphic video of murder recorded by a drug cartel.

~ (Casey Newton, on content moderation at Facebook)[1]

Facebook and Ethnic Cleansing in Burma

In 2017, the military forces in Myanmar escalated a brutal crackdown against the Rohingya, a Muslim ethnic minority living in the western region of the country.[2] About 9,000 Rohingya were murdered by military forces and close to a million fled across the border to Bangladesh.[3] About three-quarters of Rohingya living in the region at the time personally witnessed a murder, one-fifth witnessed a mass killing of more than 100 people, and the majority witnessed military forces using sexual violence against Rohingya women as part of a widespread and systematic campaign of ethnic cleansing.[4]

The military forces who perpetrated this violence toppled the democratically elected government in February of 2021, and they continue their military operation as well as their crackdown on dissent and free speech in Myanmar.[5] The United States has officially labeled the violence against the Rohingya a genocide, and the military government of Myanmar has refused to cooperate with an investigation by the International Criminal Court.[6]

Beginning in about 2016, there was an uptick of hate speech against the Rohingya on Facebook, with much of it linked to accounts run by the military forces in Myanmar.[7] Much of this speech resembled the incitements to violence that have been seen in previous genocides, including in Rwanda in 1994.[8] There were thousands of posts that promoted the dehumanization of Rohingya Muslims and incited violence against them, including "likening the Rohingya to animals, calling for Rohingya to be killed, describing the Rohingya as foreign invaders, and falsely accusing Rohingya of heinous crimes."[9] Other posts on Facebook directly incited murder, rape, and the forced displacement of Rohingya.[10] The evidence was clear: even Facebook's own head of cybersecurity policy, Nathaniel Gleicher, admitted that Facebook promoted posts that were "clear and deliberate attempts to covertly spread propaganda that were directly linked to the Myanmar military."[11]

DOI: 10.1201/9781003293125-7

The military ran accounts and spread propaganda specifically designed to inflame long-standing ethnic hatred against the Rohingya, including "sham photos of corpses that they said were evidence of Rohingya-perpetrated massacres."[12] All of this was intended to justify a military crackdown against the ethnic minority and drive them from Myanmar – which it accomplished with all-too-much success.[13]

Old Hate and New Tech

Facebook's role in the genocide in Myanmar has since been condemned by the United Nations.[14] It has also been the subject of a class action lawsuit brought by a group of Rohingya Muslims against Meta, Facebook's parent company, in California. The complaint asks for $150 billion U.S. dollars as compensation and punitive damages.[15] The lawsuit alleges that Facebook was negligent in failing to remove hate speech that incited ethnic violence but also alleges a novel claim of product liability – for the faulty design of Facebook's content moderation algorithms.[16]

Moderating content is one of the more difficult tasks that social media companies must accomplish.[17] Facebook moderates billions of posts a day,[18] in almost every language and culture around the world. They must take account of dynamic linguistic and cultural nuances in a shifting cultural sphere – one that is itself shaped by social media and its content moderation practices. Facebook has been removing more and more hate speech – about 7 million posts in the third quarter of 2019; over 80% of the hate speech removed was identified by artificial intelligence (AI) algorithms.[19] The rest are flagged by users themselves.[20] Facebook states that the final decision to remove a post for hate speech is always made by a human moderator, who reviews the algorithm's identification.[21]

Facebook has developed algorithms to identify hate speech in over 40 languages.[22] At the time of the Rohingya genocide, they did not employ any AI or human content moderators to identify and remove hate speech in any of the languages spoken in Burma.[23] Facebook is also short of human content moderators who are fluent in the languages and cultural practices of many developing countries – including many weak states in which ethnic and political violence is rife. This means that hate speech and incitement to further violence can spread faster in these countries than they do in places such as Europe and North America, creating systemic disparities in who is subjected to hate speech.[24] Facebook has since attempted to remedy this: it has now developed content moderation algorithms in Burmese and hired about 100 Burmese-speaking content moderators (for a linguistically and ethnically diverse nation of over 50 million people) who are assisting in developing better training data for Facebook's hate-speech classifiers.[25]

When one thinks of content moderation, one normally thinks of algorithms designed to scrape through massive amounts of online materials and remove a subset of offending content. But this neglects the ubiquitous algorithms that promote, recommend, and downgrade virtually *all* materials posted on all social media sites, in a never-ending quest to drive engagement (and advertising dollars). Even objectional material can be welcome when it promotes this goal – but not when it drives users away. One content moderator for Cognizant, a subcontractor for Facebook, told Casey Newton that the work of content moderators was fundamental to Facebook's brand, stating "If we weren't there doing that job, Facebook would be so ugly ... We're seeing all that stuff on their behalf."[26]

The truth is that all content is moderated all the time. The complaint against Meta attempts to capture this by alleging that Facebook's algorithms use marketing, psychology, and social science to exploit our vulnerability to emotional, sensational, and politically divisive content for the sole purpose of feeding us more and more of it.[27] As Rose-Stockwell states, "this is what social media does to us regularly: it encourages us to observe conflicts and pick sides on topics about which we would otherwise have few opinions. At its core, it is an opinion-serving machine. And on social media, not all opinions are served equally."[28] Hatred, violence – even 'fake news' and misinformation of all kinds[29] – have a significant advantage when it comes to what we engage with on social media.[30] On the other hand, the more we encourage the filtering of content, the more we infringe on the free flow of speech and ideas, and the more we promote the very 'filter bubbles' that give rise to the problem in the first place.[31]

Old hate and new tech may have clashed in a particularly toxic and dangerous fashion in Myanmar because of the relative novelty of social media among the population, combined with Myanmar's strict censorship and the paucity of sources of information. In 2014, less than 1% people had internet access, and this number grew to about 15 million by 2018 – over one-quarter of the population.[32] This happened because cheap smartphones with $1 SIM cards flooded the Burmese market after 2014 – and almost every one of these phones came with Facebook pre-installed.[33] As the human rights assessment of Facebook's role in the Rohingya genocide stated, both digital literacy and the rule of law in Burma were very weak.[34]

Myanmar is known to engage in "severe restrictions on free expression," including arbitrary detentions of journalists, and repressive criminal libel laws designed to stifle dissent.[35] Hogan and Safi report that one cybersecurity analyst in Yangon stated that this led to a situation in which "Facebook is arguably the only source of information online for the majority in Myanmar."[36] Facebook knew that the population was digitally naïve, that the government engaged in severe repression of information, and that the political climate was rife with ethnic discord and highly susceptible to hateful and violent speech. They took advantage of this situation to foster engagement, and drive growth in advertising revenues in Burma.[37] As the lawsuit against Facebook states, "Facebook made a corporate decision to lean into the hate."[38]

Whether or not the class action lawsuit against Meta is successful, it is part of a tipping point in how we view AI systems for content moderation – in how we recognize and respond to the design flaw at the heart of these algorithms. Engagement drives the social media business model because engagement means likes, shares, and therefore advertising revenues.[39] Posts with higher engagement are placed higher in social media news feeds.[40] Hateful and violent content is generated and promoted by large numbers of fake accounts, which generates high engagement, and so Facebook's algorithms "prioritize it in the News Feeds of real users."[41] In Burma, the complaint alleges that Facebook's algorithms not only failed to detect and remove hate speech against the Rohingya, but that it exploited and promoted it in users' news feeds; this had the effect of driving users' radicalization and "causing them to tolerate, support, and even participate in the persecution of and ethnic violence" against Rohingya Muslims.[42]

Commentary

Buddhist Ethics

By Peter Hershock

Computationally mediated digital connectivity makes possible the progressive conversion of attention-transmitted data into both revenue streams and powers to predict and produce human thought and behavior. These potentials range from accelerating fast fashion consumption to fostering populist political alliances, influencing swing voters, and fomenting ethnic violence. The range of these potentials from the frivolous to the murderous serves as evidence that the ethics of content moderation is not and cannot be simple.

Illustratively, this case study makes clear that Facebook served simultaneously in Myanmar as an apolitical, revenue-generating matchmaking service for producers and consumers, as a platform for inciting ethnic violence, as a conduit for sharing evidence of government and military complicity in that violence, and as a medium through which to organize protests and contest the weaponization of feelings of personal and communal precarity.

The attempt to hold Facebook liable for the violence suffered by the Rohingya rests on a readily understood logic. Facebook acted in its own commercial self-interest and in either ignorance or active disregard of the tragic potential of neglecting to monitor the content circulated through its platform. It shirked its moral responsibilities.

But do corporations have moral responsibilities? Although corporations are considered "legal persons," they are not traditional moral agents. Facebook is an enterprise for commercializing a digital medium or field of communication. It may be reasonable to require such an enterprise to assume global technical responsibility for content moderation. But assigning it local ethical responsibility is arguably another matter since the norms of desirable and permissible moderation vary so significantly among its 1.62 billion daily users. An argument can be made that, just as it is farmers – and not their fields – that determine which crops to grow and sell, it is not Facebook that is responsible for the seeds of violence sown on its platform; it is those who post hate speech.

Then again, farmers respond to market demands, and the analogy suggests that responsibility for circulating unblocked hate speech on Facebook cannot be attributed solely to those who authored hateful posts. Without the actions of those who "liked" and "shared" them, those posts could not have had such a widespread and violent consequences. The agency mediated by social media platforms presupposes, at the very least, distributed responsibility.

Matters become more complex still when it is considered that, while all of those involved as direct or indirect agents of the violence and disruptions of daily life that resulted from Facebook-circulated hate speech, they also became the patients of those communicative actions and all that followed from them, including international sanctions, reputational harms, and moral condemnation. In sum, the global infrastructure of digital connectivity not only renders national borders porous; it blurs the conceptual boundaries among ethical agents, actions, and patients.

AI is often referred to as a general-purpose technology and is generally presumed to be ethically neutral. Any harms resulting from intelligent technology are the responsibility of those who design, deploy and use algorithmic and deep learning tools. That is, these harms are taken to be functions of either accident of design or misuse by design. Buddhist relational ontology suggests otherwise.

To begin seeing why, it is first useful to distinguish between tools and technologies. *Tools* are localizable artifacts that extend or augment our capacities for action, and that we are individually free to use or not. *Technologies* are non-localizable relational media that scale up human intentions and values, selectively altering the environments within which we make decisions and act, affecting both how and why we do so. That is, technologies emerge from and inform/structure human conduct much as ecosystems emerge from and recursively inform/structure species relationships. We do not have "exit rights" from technology.

Accidents of design and misuse by design are tool risks. The harms made possible by Facebook's limited or faulty content analyzing algorithms are an example of the former. Using Facebook to promulgate hate speech and incite ethnic violence are examples of the latter. Both are fundamentally agent-originated risks. Technological risks are structural and relational. Unlike tool risks, they originate in the complex and recursive conditioning of human-technology-world relations. Technological risks include much more than the local harms that occur at the final causal stages of tool use – the point at which agents put intentions into action, with or without due consideration of the patients of that action. Thus, when gun rights lobbyists argue that "guns don't kill; people do," they are performing conceptual sleight-of-hand — skillfully misdirecting critical attention toward tools (guns) and their designers and users, and away from weapons technology – a relational medium that restructures decision-making environments in ways conducive to rationalizing causing harm from a distance in response to perceived threats, insults, or collisions of interest.

Buddhism ethics, based on recognition of the interdependent origination of all things, suggests that it is less crucial to ask which agents are responsible for certain outcomes than it is to ask what values and intentions are at play in shaping the circulatory dynamics of social media. What patterns of experiential outcomes and volitional opportunities are being scaled up and recursively reinforced by the relational media of intelligent technology?

Reflecting its American origins and Silicon Valley liberalism, Facebook valorizes freedoms of choice and speech and presumes the wisdom of minimal and purely reactive content moderation. The ramifications of social media use in Myanmar call this into question. In contrast, while China uses the same AI and human screening techniques as Facebook, its digital content curation policies and practices are centered on the values of political stability and social harmony. Moreover, while its content moderation still aims at harm prevention, it also aims to produce good citizenship habits through conduct incentives. Content curation is proactively oriented to promote social well-being, as defined by the party state.

From a Buddhist perspective, whether China's proactive approach to content curation or Facebook's minimalist and reactive content moderation approach, is to be lauded or lamented, should not rest on approving or disapproving their uses of machine learning tools, their economies of responsibility, or even their short-term impacts on individual social media users. It should depend on their medium and long-term relational ramifications and risks.

Some guidance for this assessment is offered by the Buddhist Eightfold Path, which charts a course toward enlightened and enlightening presence through cultivating right or corrective views, intentions, speech, conduct, livelihood, effort, mindfulness, and concentration. Traditionally, right/corrective speech involves refraining from lying, backbiting, slander, harsh, or abusive language, as well as frivolous chatter and gossip. Much of what is circulated by social media would clearly not qualify.

A more positive characterization of faring well in Buddhist practice is that it results in suffusing one's entire situation with the relational qualities of compassion, lovingkindness, equanimity, and joy in the good fortune of others. At present, these are not the objective functions that AI systems are currently directed to optimize. But could they be?

Facebook has successfully experimented with manipulating social media feeds to affect user emotions. Pilots of China's social credit system have proved viable means of fostering public civility, and Japanese researchers have developed AI systems that accurately read emotions and engage humans in ways that are experienced as caring. All of this suggests that there is no technical barrier to establishing compassion, lovingkindness, equanimity, and sympathetic joy as objective functions of AI-enacted social media content curation. Digital connectivity could, perhaps, be liberating by design.

Virtue Ethics

By John Hacker-Wright

This case brings out the truly dire consequences that can result from the new possibilities for the manipulation of public opinion through social media. There are obvious failings in the corporate leadership of Meta which exhibited a form of injustice in privileging profits above safety of the Rohingya targeted by hate speech.

In this comment I will start from the rather pessimistic stance that despite the increased efforts on the part of social media companies, we are not likely to see an end to hate speech and other forms of manipulative content on social media platforms. After all, as pointed out in the chapter, there is a strong incentive for social media companies to keep and promote inflammatory content given that it furthers engagement. But even aside from that, it is unrealistic to think that all such content could be identified and removed even with the best of intentions and robust funding. Given that, what virtues might we develop as users that would protect against such manipulation and potentially save lives?

It is important to note that susceptibility to such manipulation trades on some good qualities that we should not sacrifice in thinking about how to inoculate ourselves against such tragic outcomes as played out in Myanmar. It is only people who are concerned about the conditions of their communities and nation that can be manipulated on the basis of that concern. Civic friendship, which consists of a shared sense of identity and mutual concern for members of one's community, and patriotism are arguably virtues insofar as in otherwise virtuous agents, they motivate members of a community to act for a common good.

Patriotism is often associated with a blind passion for one's country that motivates unthinking loyalty, narrow jingoistic concern, and mistreatment of outsiders. If patriotism is used as a virtue term, it is not a passion, but a discerning trait of character that regulates our love of our country in accordance with its qualities, something like a proper pride taken in one's own actions that accords with their actual merits. Proper patriotic pride could be taken in helping neighboring countries, welcoming newcomers, and resisting demagoguery. Patriotism would also motivate feeling ashamed when one's country acts unjustly. Hence, it is plausible to treat patriotism as a virtue term, while acknowledging that the term is often not used in that way.

But otherwise good qualities such as civic friendship and patriotism can be exploited to motivate bad actions through misinformation. Someone who is deeply concerned about their community, when misinformed about a threat to their community, may act unjustly against the perceived threat. Using lethal force to defend one's community against a threat is generally considered morally permissible. The medieval philosopher and theologian Thomas Aquinas, who elaborated and systematized Aristotle's ethical theory, developed a theory of just war according to which a threat to one's community is clearly a sufficient condition for entering into military conflict (though not for committing atrocities against one's enemy once the combat has begun). Yet Aquinas does not make explicit what is obviously assumed as a background in his theory: the judgment that there is a threat needs to be epistemically sound.

In this regard, the virtue of practical wisdom is crucial. Practical wisdom is an intellectual virtue that regulates our thinking about what to do. Someone with practical wisdom deliberates well about action. A significant part of reasoning well about what do involves reasoning from true premises, and attaining true premises, is not easier in practical life than in scientific contexts, but it involves distinctive challenges. One aspect of reasoning from true premises, according to Aristotelian virtue ethics, involves having moral virtues like courage, which regulates our passions of fear and confidence, and temperance, which regulates our appetites for food and sex.

Consider this: for someone who lacks courage, a threat appears to be much greater than it actually is – their fear is disproportionate to the object. To someone who is intemperate, food that is unwholesome or belongs to someone else will appear to be good to eat. Here we see a basic Aristotelian commitment: in practice our judgments about the world we are situated in are based on our feelings. Lack of virtue, then, distorts our perception and thereby gives rise to false premises, e.g., "that man, over there, is exceedingly dangerous," or "that cupcake is good to eat." From those false premises, we are likely to draw conclusions that lead to bad action, like "I will run away" or "I'll eat it." In the case of anger, someone who is quick to anger might falsely judge someone worthy of retribution. Of course, our reasoning is not always spelled out so explicitly, and often occurs in a fast, implicit, System 1 way, but if spelled out, it might look something like the reconstruction. The upshot is that in the absence of moral virtues, our reasoning about how to act will be distorted. Hence, it is important to have virtue in order to respond well to the situations that we find ourselves in, including situations that are presented to us through social media.

So far, the picture of practical reasoning I have presented is largely 'perceptual', which is to say it presents us as we look out at the world from the standpoint of our desires, and if our desires are in order, through a good upbringing, we will tend to act well. But there is more to possessing practical wisdom than having moral virtues. According to Aristotle, people with practical wisdom "have a theoretical grasp on what is good for themselves and for human beings."[43] This is partly a matter of having some basic factual knowledge of what is good in domains like nutrition, but far more importantly a matter of having insight into what is good in terms of a well-lived life; in other words, insight into what actions are best for a human being. Aristotle thought that Pericles, a famous statesman who led Athens through a portion of the Peloponnesian War, was such a man. It is not entirely clear what Aristotle was praising in upholding Pericles, but in a famous funeral oration, as reported by Thucydides, Pericles praises a life of active community engagement, pursuing honor, and risking death for the sake of freedom.[44] Even if we disagree with this conception of the human good, it is against a background of some general views about what constitutes the best sorts of activities for a human being that particular proposals will be assessed, and these general claims are also among the premises of our reasoning, "it befits a human to ..." or "it is best for humans to..."

A further component of practical wisdom, which is especially important to this case, is an intellectual capacity that Aristotle calls 'comprehension', which is a capacity by which we understand the situation we are in. In explicating this quality, the Aristotelian virtue ethicist Rosalind Hursthouse points out the following:

> A 'situation' which calls for my doing something may not be facing me at all, waiting for me to read it, but rather something whose details I have to work out from what other people say about it. And until I can make a correct judgement about their accounts of the relevant matters, any practical conclusion I reached about what to do in 'this situation' would be made in the dark.[45]

Hursthouse's account of this component of practical wisdom acknowledges it as a fact about human life that we are very often in the situation of building a picture of what is happening in our community on the basis of other people's accounts of it. And very often others are misinformed or aiming to deliberately disinform. We must acquire some degree of savvy in discerning who is trustworthy. Such accounts create the appearance of a dilemma or necessary paths of action where there are more options. Assessing the accounts and being able to reject them completely and do more investigating on one's own is a capacity crucial to practical wisdom that takes experience to develop.

What we see in this case is the need for a 'technomoral' extension of Aristotle's notion of comprehension.[46] It is obvious that mass media and social media create new challenges for comprehension that require us to develop appropriate skills for recognizing distortions. We can come to recognize political bias and sensationalism in mass media and put breaks on our inclinations to act on their basis. Likewise, we can recognize the tendency of social media to feed us what we are already inclined to believe, to close out countervailing views that might challenge our judgments, and to harbor fake bot accounts that guide us to conclusions, we would not otherwise reach.

In the context of emerging technologies, the component of practical wisdom, that Aristotle spent only a few lines of his treatise on, takes on an outsized importance. Technomoral comprehension as a capacity to discern reliable information on social media is a capacity that is, I think, not widely exhibited even in places that have had social media from its inception. Its development will surely be an uneven process, but it is equally sure the crucial virtue that we need in the face of widespread manipulation on social media platforms. And it is not something that can be designed into the architecture of social media. If anything, overconfidence in algorithms will impede its development. Flags placed on stories deemed suspicious may lure us into misplaced confidence in stories not so flagged. If the designers of social media platforms want to help, it is perhaps best done by encouraging and funding independent, educator-led efforts to build critical thinking media skills that bolster technomoral comprehension.

Indigenous Ethics

By Joey Miller and Andrea Sullivan-Clarke

The connection between content moderation and the ethnic cleansing in Myanmar (i.e., Burma) demonstrates the moral importance of the idea in Indigenous philosophy that everything is connected. Algorithms on their own do not cause harm. However, how they are designed and how they are used can most certainly cause harm. This is because of a myriad of facts concerning how humans relate to, and interact with, their surroundings (e.g., the various psychological, biological, social, physiological, etc., features of humans and how these features are used to navigate one's surroundings). That means there are moral implications for how algorithms are designed and used. While Facebook itself did not directly or intentionally contribute to the murders and fleeing of the Rohingya, Facebook undoubtedly bears some responsibility in the genocide.

On an Indigenous understanding of ethics, Facebook's responsibility does not result from them not employing enough people who spoke the languages of Burma. As such, hiring more Burmese-speaking content moderators does not fix the problem. Rather, Facebook's utilizing algorithms that are designed in such a way as to promote engagement by exploiting humans' psychological (i.e., biological, social, etc.) features is problematic. More specifically, there is a lack of foresight or consideration concerning the consequences of the algorithms Facebook uses, which means the connection between these algorithms and their effects – i.e., the interconnectedness of all things – are either not considered, ignored, or overlooked. Actions are not recognized as (inter)actions.

It is clear, at the very least, that not enough consideration went into the harms that might result from these kinds of algorithms. Given the increased harm that resulted from enabling the spread of misinformation, as well as promoting hate speech and violence, it is clear that Facebook did not have a sufficient understanding of their clientele or the consequences of their algorithms. In this sense, Facebook contributed to upsetting (or at least further upsetting) the balance and well-being among the people and groups involved in the genocide in Myanmar. In other words, Facebook contributed to the promotion of disharmony.

Given that, in Indigenous philosophy, (inter)actions are right or wrong to the extent that they promote or disrupt harmony,[47] it is clear that Facebook's use of *these* algorithms is wrong. However, "indirect" or "unintentional" Facebook's practices were, they still contributed to the disruption of harmony. Even if disharmony already existed in Myanmar, further disrupting harmony is still wrong. This does not necessarily rule out the use of all algorithms for the purpose of content moderation. It just means that *if* algorithms are going to be used, the persons using them ought to have a sufficient understanding of their limits, reception, and consequences. Without such an understanding, there is a greater tendency to disrupt harmony.

There is no universally applicable rule regarding how harmony ought to be promoted in all situations.[48] While there is a general idea that harmony ought to be promoted, *how* harmony is promoted depends on one's surroundings (i.e., their situation and circumstance).[49] Part of the motivation behind this idea is that there are too many variables to account for when designing ethical theories. This is also true of designing algorithms. Algorithms, like moral theories, cannot possibly be designed to account for all possible instances of wrong. In fact, also like moral theories, they can be used to promote wrong. These two features or algorithms, (1) their limits and (2) their being used to promote wrong, can also help to emphasize why, in Indigenous philosophy, there is no ethical principle that is applied the same in all situations.

Circumstances, situations, and context matters. Expressing intellectual humility by being appropriately aware of the limits of an algorithm and how they can be used for wrong should help to guide how one designs algorithms, or which algorithms one decides to use. In Indigenous philosophy, words have power.[50] By relegating the moderation (e.g., the promotion and regulation) of words to algorithms, one fails to either recognize the limits of such algorithms and/or the power of words. In other words, they fail to display humility, and this tends to lead to a disruption of harmony or the promotion of disharmony.

Notes

1. Casey Newton, "The Trauma Floor," *The Verge*, February 25, 2019, https://www.theverge.com/2019/2/25/18229714/cognizant-facebook-content-moderator-interviews-trauma-working-conditions-arizona.
2. Connor Finnegan, "US Recognizes Myanmar's Atrocities Against Rohingya as 'Genocide'," *ABC News*, March 21, 2022, accessed March 30, 2022, https://abcnews.go.com/Politics/us-recognizes-myanmars-atrocities-rohingya-genocide/story?id=83582997.
3. Finnegan, "US Recognizes Myanmar's Atrocities."
4. Finnegan, "US Recognizes Myanmar's Atrocities."
5. United States Department of State, "2020 Country Reports on Human Rights Practices: Burma," United States Department of State, Bureau of Democracy, Human Rights and Labor Report March 31, 2020, https://www.state.gov/reports/2020-country-reports-on-human-rights-practices/burma/.
6. Finnegan, "US Recognizes Myanmar's Atrocities."
7. Libby Hogan and Michael Safi, "Revealed: Facebook Hate Speech Exploded in Myanmar During Rohingya Crisis," *The Guardian*, April 2, 2018, https://www.theguardian.com/world/2018/apr/03/revealed-facebook-hate-speech-exploded-in-myanmar-during-rohingya-crisis.

8. Allan Thompson, ed. *The Media and the Rwanda Genocide* (Kampala, Uganda: Fountain Press, 2007).
9. *Jane Doe* v. *Meta Platforms Inc. Class Action Complaint*, Superior Court of the State of California for the County of San Mateo, n.d., para. 20.
10. Paul Mozur, "A Genocide Incited on Facebook, With Posts from Myanmar's Military," *New York Times*, October 15, 2018, https://www.nytimes.com/2018/10/15/technology/myanmar-facebook-genocide.html.
11. Mozur, "Genocide Incited on Facebook."
12. Mozur, "Genocide Incited on Facebook."
13. Mozur, "Genocide Incited on Facebook."
14. Matthew Ingram, "Facebook Slammed by UN for its Role in Myanmar Genocide," *Columbia Journalism Review*, November 8, 2018, https://www.cjr.org/the_media_today/facebook-un-myanmar-genocide.php.
15. *Jane Doe* v. *Meta Platforms Inc. Class Action Complaint*, Superior Court of the State of California for the County of San Mateo, n.d., 1.
16. *Jane Doe* v. *Meta Platforms*, para. 41.
17. Jason Koebler, and Joseph Cox, "Inside Facebook's Struggle to Moderate Two Billion People," *Motherboard*, August 23, 2018, https://www.vice.com/en/article/xwk9zd/how-facebook-content-moderation-works.
18. Newton, "The Trauma Floor."
19. Billy Perrigo, "Facebook Says It's Removing More Hate Speech Than Ever Before, but There's a Catch," *Time*, November 27, 2019, https://time.com/5739688/facebook-hate-speech-languages/.
20. Perrigo, "Facebook Says."
21. Perrigo, "Facebook Says."
22. Perrigo, "Facebook Says."
23. Perrigo, "Facebook Says."
24. Perrigo, "Facebook Says."
25. Perrigo, "Facebook Says."
26. Newton, "The Trauma Floor."
27. *Jane Doe* v. *Meta Platforms*, para. 41.
28. Tobias Rose-Stockwell, "Facebook's Problems Can be Solved with Design," *Quartz*, April 30, 2018, https://qz.com/1264547/facebooks-problems-can-be-solved-with-design/.
29. Soresh Vosoughi, Deb Roy, and Sinan Aral, "The Spread of True and False News Online," *Science,* 359, no. 6380 (2018): 1146–1151, doi: 10.1126/science.aap9559.
30. Steve Rathje, Jay J. Van Bagel, and Sander van der Linden, "Out-Group Animosity Drives Engagement on Social Media," *Proceedings of the National Academy of Sciences*, 26, June 29, 2021, doi: https://doi.org/10.1073/pnas.2024292118.
31. Rose-Stockwell, "Facebook's Problems."
32. Libby Hogan, and Michael Safi, "Revealed: Facebook Hate Speech Exploded in Myanmar During Rohingya Crisis," *The Guardian*, April 2, 2018, https://www.theguardian.com/world/2018/apr/03/revealed-facebook-hate-speech-exploded-in-myanmar-during-rohingya-crisis.
33. Ingram, "In Some Countries."
34. Business for Social Responsibility, *Human Rights Impact Assessment: Facebook in Myanmar*, Report, October 2018, 3.
35. United States Department of State, "2020 Country Reports on Human Rights Practices: Burma."
36. Hogan and Michael Safi, "Revealed: Facebook Hate Speech."
37. *Jane Doe* v. *Meta Platforms*, para. 10.
38. *Jane Doe* v. *Meta Platforms*, para. 11.
39. *Jane Doe* v. *Meta Platforms*, para. 36.
40. *Jane Doe* v. *Meta Platforms*, para. 38.
41. *Jane Doe* v. *Meta Platforms*, para. 17.

42. *Jane Doe* v. *Meta Platforms*, para. 168.
43. *Nicomachean Ethics*, trans. C.D.C. Reeve (Indianapolis: Hackett Publishing, 2014), 102, 1140b8–9.
44. Robert B. Strassler, ed. *The Landmark Thucydides* (New York: Free Press, 1996), 115–118.
45. Rosalind Hursthouse, "Practical Wisdom: A Mundane Account." *Proceedings of the Aristotelian Society*, 106 (2006): 293.
46. 'Technomoral' is Shannon Vallor's term, a modifier for virtues that we need to develop in the light of emergent technologies. See: *Technology and the Virtue* (Oxford: Oxford University Press, 2016), 10.
47. Jean Chaudhuri, Jean, and Joyotpaul Chaudhuri, *A Sacred Path: The Way of the Muscogee Creeks* (Los Angeles: UCLA American Indian Studies Center, 2001); Viola F. Cordova, *How It Is: The Native American Philosophy of V.F. Cordova*, ed. Kathleen Dean Moore, Kurt Peters, Ted Jojola, Amber Lacy, with Foreword by Linda Hogan (Tucson: University of Arizona Press, 2007); Joseph Len Miller, "Etemeyaske Vpokat (Living Together Peacefully): How the Muscogee Concept of Harmony Can Provide a Structure to Morality," in *Comparative Metaethics: Neglected Perspectives on the Foundations of Morality*, ed. Colin Marshall (New York, NY: Routledge, 2020).
48. c.f., Joseph Len Miller, "What Do We Need to Know to Live in Harmony with Our Surroundings?" in *Ways of Being in the World: An Introduction to Indigenous Philosophy*, ed. Andrea Sullivan-Clarke (Peterborough: Broadview Press, Forthcoming); Miller, "Etemeyaske Vpokat."
49. Though not discussing harmony in particular, the idea that ethics is grounded in one's surroundings – particularly the land – is discussed in detail in Brian Burkhart, *Indigenizing Philosophy through the Land: A Trickster Methodology for Decolonizing Environmental Ethics and Indigenous Futures* (East Lansing: Michigan State University Press, 2019).
50. "One of the greatest lessons the ancestors have given us is this: what we speak, we create. Our words are powerful, and they can heal as well as destroy, which is why we must be mindful of the intention behind our words and use our voice for good," Doug Good Feather, *Think Indigenous: Native American Spirituality for a Modern World* (New York City: Hay House, Inc., 2021), 38.

8 Mental Malware

Algorithms and Choice Architecture

> We don't want to ask people what they're going to do ... because we know that's not very predictive of how an advert is going to perform, because people go into their left brain and start thinking too much.
>
> ~(Kerry Collinge, marketing executive at the 'System 1 Group' marketing firm)[1]

The Cambridge Analytica Data Scandal

The Cambridge Analytica data scandal had its roots in 2010 when Facebook launched their 'Open Graph' application.[2] Open Graph permitted the developers of third-party apps to request access to Facebook user's personal data, as well as all of their 'friends' data.[3] In 2013, academic researcher Alexsandr Kogan, in conjunction with marketing and data analytics firm Cambridge Analytica, launched an app called 'this is your digital life'.[4] The app invited users to take a free personality quiz, and about 300,000 Facebook users did so.[5] The app collected data about their psychometric profiles from the quiz – which measured users' 'big five' personality traits[6] – but also freely harvested Facebook data from all of their friends.[7] Cambridge Analytica was trying to amass a dataset on as many U.S. voters as possible.

In 2015, the first reports came in that the political campaign of Ted Cruz had analyzed millions of these psychometric profiles in an effort to gain an advantage in his election to the U.S. Senate – a revelation that proved quite unpopular and led to assurances from Facebook and Cambridge Analytica that the data in question had been deleted.[8]

However, in 2018 the news broke that the Facebook data of tens of millions of users – perhaps as many as 87 million – was harvested by Cambridge Analytica and used in the 2016 U.S. Presidential election by Donald Trump's campaign.[9] Many of these individuals had *not* taken the personality test but, since a friend had, researchers could freely access their data.[10] Cambridge Analytica then cross-referenced the Facebook data with other data they had purchased as well as local electoral rolls.[11] Cambridge Analytica was thus able to compile extensive dossiers on tens of millions of voters, including their demographic characteristics, personality traits, social networks, purchasing

DOI: 10.1201/9781003293125-8

history, likes, political party membership, etc.[12] McNamee estimates that these dossiers ended up including about 13% of all eligible U.S. voters.[13]

Cambridge Analytica used their voter profiles to gain an advantage in the 2016 election of Donald Trump as President of the United States, as well as the 'leave' campaign of the Brexit referendum in the UK.[14] As Cadwalladr explains, Cambridge Analytica "used the test results and Facebook data to build an algorithm that could analyze individual Facebook profiles and determine personality traits linked to voting behaviour."[15] This algorithm was particularly effective because it allowed data scientists to identify swing voters, and then to target them with specific ads and messages that were most likely to 'nudge' their vote.[16]

Choice Architecture and Persuasive Technology: A 'Skinner Box for the Modern Human'

Cambridge Analytica used the data to train recommendation algorithms and promote social media content "capable of moving public opinion at scale."[17] Cambridge Analytica's algorithms did this by micro-sorting people into groups defined by demographic, political, and psychometric characteristics: for example, socially conservative female voters in affluent suburbs of D.C. whose children were affected by COVID-related school closures, working class voters in the rust belt who are long-term underemployed, lower-class retirees from Central Florida who are concerned about rising health care costs. These algorithms are intended to target the much-coveted 'swing voters' so that they can influence their vote in crucial battleground districts.[18] More than this, they can train and improve their algorithms in real time on focus groups.[19]

Although Facebook was fined 5 billion U.S. dollars by the Federal Trade Commission over the data breach,[20] the dangers of what is known as 'persuasive technology' go far beyond the Cambridge Analytica data scandal. Educators and psychologists have been raising concerns about persuasive tech for years, but this had little impact on the consumer and political marketing industries.[21] The scandal is about far more than the breach of users' privacy. As McNamee states, it is about how our "data is feeding artificial intelligences whose objective is to manipulate the attention and behavior of users without their knowledge or approval."[22]

While marketing based on demographic groupings has a long history, the practice has been supercharged by the enormous amounts of data created by social media. Facebook, for example, has created groups called 'lookalikes' which classify users into groups with similar profiles to assist the company in microtargeting them.[23] Christopher Wylie, a former data scientist turned whistle-blower at Cambridge Analytica, states that users are served up with content based on their lookalike group that other users do not see; this has the effect of creating filter-bubbles and deepening social divisions.[24] He states that there is "a fine line between an algorithm defining you in order to represent *who you really are* and an algorithm defining you to create a self-fulfilling prophesy of *who it thinks you should become*.[25]

Lookalikes and microtargeting harness the power of data science to engage – albeit much more effectively – in the age-old practice of propagandizing. Marketers, for consumer corporations and political campaigns, routinely serve up moralized and highly emotive content, which spreads particularly rapidly on social media.[26] In fact, social media and other online platforms are now the "primary sources of morally relevant stimuli people experience in their daily life."[27] Those who would leverage social media to influence public opinion also make use of social reward learning – usually in the form of 'shares', 'clicks', 'likes', 'followers' and other reinforcing forms of engagement. These behaviors are not only highly rewarding, but they can also be mined by machine learning systems to predict our *future* behavior, our friends and followers, and our lookalike groups.[28] Lindstrom et al. conceptualized an individual engaging with online content as being like a mouse in a Skinner box pushing a reward lever,[29] and they concluded that social media is like "a Skinner box for the modern human."[30]

The comparison between social media engagement and an experimental mouse in a box goes deep. In 2014, a study was published involving a collaboration between Facebook and Cornell University – this experiment involved not the Food & Brand Lab but the Department of Communication & Information Science.[31] Researchers manipulated the emotional content of posts that Facebook users were receiving in their News Feeds, especially from people they trusted, such as friends and those they followed.[32] The researchers wanted to see if positive versus negative emotional content would affect the mood of users' subsequent postings – i.e., whether there was evidence of 'emotional contagion' operating on social media.[33] There was,[34] but this was not the most significant aspect of the study. The participants were not informed that they were being used as research subjects; indeed, their consent to participate in experimental research was never sought.[35] Since the data was collected by Facebook, researchers did not even seek the approval of Cornell's research ethics review board.[36] It is not clear that they would have received approval if they had.

When news of the lack of consent broke, there was a backlash against the study.[37] The core principle of research ethics requires obtaining informed consent from research participants. This principle was established in the field of medical ethics (see Box 8.1) but has since spread to all fields of academic research involving human subjects.[38] Many bioethicists argued that the research did not "egregiously" breach any principles of law or ethics, and if it did then it meant that Facebook's standard practices were also ethically dubious.[39] Catherine Flick responded by arguing that the ethics of corporations routinely subjecting users to experimental and experimental-like manipulation without their knowledge or consent was precisely what was in issue.[40]

While the lead scientist in the emotional contagion study apologized, this ethical breach foreshadowed the Cambridge Analytica scandal that

---

Box 8.1

The Principles of the Nuremberg Code

The Nuremberg Principles regarding human experimentation are:

1   The voluntary consent of the human subject is absolutely essential.
2   The experiment should be such as to yield fruitful results for the good of society, unprocurable by other methods or means of study, and not random and unnecessary in nature.
3   The experiment should be so designed and based on the results of animal experimentation and a knowledge of the natural history of the disease or other problem under study that the anticipated results will justify the performance of the experiment.
4   The experiment should be so conducted as to avoid all unnecessary physical and mental suffering and injury.
5   No experiment should be conducted where there is an a priori reason to believe that death or disabling injury will occur; except, perhaps, in those experiments where the experimental physicians also serve as subjects.
6   The degree of risk to be taken should never exceed that determined by the humanitarian importance of the problem to be solved by the experiment.
7   Proper preparations should be made and adequate facilities provided to protect the experimental subject against even remote possibilities of injury, disability, or death.
8   The experiment should be conducted only by scientifically qualified persons. The highest degree of skill and care should be required through all stages of the experiment of those who conduct or engage in the experiment.
9   During the course of the experiment the human subject should be at liberty to bring the experiment to an end if he has reached the physical or mental state where continuation of the experiment seems to him to be impossible.
10 During the course of the experiment the scientist in charge must be prepared to terminate the experiment at any stage, if he has probably cause to believe, in the exercise of the good faith, superior skill, and careful judgment required of him that a continuation of the experiment is likely to result in injury, disability, or death to the experimental subject.

---

would shortly follow.[41] Prominent professors at prestigious universities like Cambridge and Harvard knew of Cambridge Analytica's algorithm development, and they found it exciting and ground-breaking. There seems to have been no discussion of research ethics.[42] As Wylie states, "with scholars from the world's leading universities telling me we were on the cusp of 'revolutionizing' social science, I had gotten greedy, ignoring the dark side of what we were doing."[43]

Persuasive tech is part of the broader field of 'choice architecture.' A choice architect organizes the "context in which people make decisions."[44] There is no such thing as a neutral context: all conditions exert at least some pressure to decide one way or another.[45] We are all novices in a complex and dynamic social sphere populated by experts, professionals, and highly trained algorithms – whose combined purpose is mainly to sell us something or to nudge us to adopt one opinion or political candidate over another.[46] Recommendation algos leverage cognitive biases and unconscious predispositions, taking advantage of "busy people trying to cope in a complex world in which they cannot afford to think deeply and at length about every choice they have to make."[47]

Marketers can look at our – unconscious and split-second – emotional reactions by analyzing our facial expressions in real time. They can then use this data to target messages and influence voting and consumer behavior.[48] This takes advantage of what psychologist Daniel Kahneman calls 'System 1' thinking – our automatic, emotive, and unconscious thinking that bypasses the reasoned and deliberate thinking of 'System 2'.[49] In this way, marketing firms like Cambridge Analytica can ensure that our choices are automatic, easily manipulated by emotional and moralizing content, and designed to make sure that we don't spend any time "thinking too much."[50]

'Popular' Politicians, 'Inauthentic' Activity and the Mathew Effect

Choice architecture and persuasive tech have been used to great effect by politicians and political campaigns worldwide, including by several authoritarian regimes. Juan Orlando Hernández, the president of Honduras, amassed hundreds of thousands of followers and likes on Facebook by creating fake pages and user profiles.[51] All these pages were run by the same person who administered Hernández's own social media accounts.[52] Hernández is a nationalistic and autocratic ruler who supported the 2009 coup in Honduras. He has been accused of manipulating his 2017 election win[53] by using tactics similar to what Russia is accused of employing in the 2016 U.S. election.[54]

Sophie Zhang, a data scientist at Facebook turned whistleblower,[55] wrote a 6,600-word memo to expose the fraud.[56] Her job had been combatting fake engagement of this type on Facebook.[57] She explains that "Sitting behind a computer screen, the administrator could publish a post about how well Hernández was doing his job on the president's Facebook Page, then use his hundreds of dummy Pages to make the post appear popular[.]" This is the "digital equivalent of bussing in a fake crowd for a speech"[58] – a type of inauthentic activity also known as 'astroturfing'.

This type of fake engagement, known as 'coordinated inauthentic behavior' in Facebook's Community Standards,[59] leverages the 'Mathew effect'. This refers to the tendency for someone who has an initial advantage in a system to accumulate more over time.[60] The name comes from the Gospel according to Mathew, which states, "For to the one who has, more will be given, and he will have an abundance, but from the one who has not, even what he has will be taken away."[61]

Mathew effects are sometimes beneficial, but they more often produce injustices.[62] They have been found to play a key role in sustaining and widening economic and social stratifications of all kinds.[63] As all complex systems are dynamic, the opposite effect does sometimes occur – once in a while the poor get richer – but this is rarer and the effects are weaker.[64] These feedback loops are part of many natural and ecological systems and living species.[65] All throughout the natural world, it seems that the rich do have a marked tendency to get richer and the poor to get poorer. Politicians who have many likes and followers get more engagement on social media, and those with fewer get less.

Facebook's threat assessment team confirmed Zhang's findings that the Honduran President was engaged in a coordinated inauthentic activity. An internal report from Facebook stated that his campaign had "persistently boosted a likely illegitimate president in an ARC [at-risk country]," and that this likely had "IRL [in-real-life] impact."[66] Nearly 1,500 pages and hundreds of accounts were taken down by July of 2019.[67]

The takedowns had a little long-term impact; when fake accounts and pages are taken down, new ones go up again the next day.[68] Nor was Hernández the only autocratic leader leveraging the Mathew effect in their social media campaigns. Zhang states that similar inauthentic activity was being employed by networks in countries around the world, including Afghanistan, Albania, Azerbaijan, Bolivia, the Dominican Republic, Ecuador, El Salvador, India, Indonesia, Iraq, Italy, Mexico, Mongolia, Paraguay, Philippines, Poland, South Korea, Taiwan, Tunisia, Turkey, and Ukraine.[69]

The inauthentic activity from Azerbaijan is particularly troubling because of the government's poor human rights record and tendency to use authoritarian sanctions and violence to repress journalists and government critics and to limit internet freedoms and access to information.[70] Wong reports that President Ilham Aliyev and his New Azerbaijan Party used fake accounts on Facebook as part of a campaign to target journalists and dissenting voices. She states that over a three-month period in 2019, "it produced approximately 2.1m negative, harassing comments on the Facebook Pages of opposition leaders and independent media outlets, accusing them of being traitors."[71]

Zhang states that Facebook deprioritized threats not directly affecting the geopolitical interests of North America and Western Europe,[72] leaving little redress for citizens suffering under autocratic rulers in countries like Honduras and Azerbaijan. As with the hate speech that Facebook permitted in Burma, abuses perpetrated in non-Western, poorer, nations were generally overlooked.[73] Those with fewer freedoms get less. Facebook has countered by claiming that it is their policy to prioritize the most urgent threats, but Zhang says that the problem is serious and Facebook is not committing enough resources to a problem that they have been instrumental in creating.[74] She states, "In the three years I've spent at Facebook, I've found multiple blatant attempts by foreign national governments to abuse our platform on vast scales to mislead their own citizenry, and caused international news on multiple occasions."[75] She also states that Facebook is not being transparent about their takedowns for coordinated inauthentic activity.[76]

Facebook is increasingly playing a leading role in shaping politics, public opinion, and public policy debates – even swaying election outcomes – around the world, and not just for the 2.8 billion members of humanity who use its services directly.[77] This gives the leadership of social media companies like Facebook an outsized and unaccountable role in global politics. Wong states that this permits some Facebook staff to "act as a kind of legislative branch in Facebook's approximation of a global government," while "others are more like a privatized diplomatic corps, staffing offices around the world to liaise with local businesses, civil society groups, government regulators and politicians."[78] Much of this power comes from non-transparent algorithms that promote, recommend, filter and mediate the content that appears (and which is *not* permitted to appear) on their platforms, and that can be manipulated by bad actors, governments, and the companies themselves.[79]

The Cyberspace Administration of China has recently taken steps to regulate algorithms to improve transparency and prevent some of these problems – and their efforts are being watched by regulators around the world.[80] These regulations are designed to prohibit discrimination by algorithms, prevent addiction and overuse, protect consumers from price gouging, and to improve transparency and user control over recommendation and filtering algos to foster "social fairness and justice."[81] They also aim to promote a "mainstream value orientation," to "actively disseminate positive energy," and prohibit misinformation that will negatively impact China's national interests or economic markets.[82] There is a very real threat that nation-states will use these algorithms to censor opposing views while boosting their own propaganda, along with a possibility that these algorithms – and those who control them – will shape global affairs in the 21st century.

The Nuremberg Code of Medical Ethics

At the war crimes tribunals at Nuremberg that followed World War II, almost two dozen doctors and scientists were tried for performing illegal scientific experiments on inmates, many of whom were Jewish, in Nazi concentration camps.[83] These experiments, including those performed on Jewish children, involved torture, mass murder, and euthanasia.[84] Known as the 'Doctors Trial,' it sat from 25 October 1946 to August 20, 1947.[85]

As a result of the Doctors Trial, a new standard was developed by which human experiments would be governed by the principles of voluntary informed consent on the part of those being experimented upon, and that experiments would be directed toward benefitting society as a whole. It has become one of the most important documents guiding scientific ethics in the modern era.[86] It continues to guide research ethics and experimentation on human subjects throughout the world. Codes of research ethics do yet not apply to experiments performed by private companies.

Commentary

Buddhist Ethics

By Peter Hershock

AI systems that are adept to predicting human likes, dislikes, emotions, choices, and actions can also be directed to produce them. In the case of Cambridge Analytica, its stated corporate purpose is to shape public opinion and choices by exploiting the capacity of machine learning systems to convert the epistemic resources generated by social media, e-commerce, and digital search into ontological power – the tactical to digitally individuate to informationally manipulate.

Moral outrage at Cambridge Analytica's "business plan" is not hard to appreciate. As the case study points out, however, there are no truly neutral choice contexts. The entire content of the internet cannot be presented simultaneously to any user, and thus values and intentions are necessarily infused into the choice architecture of digital connectivity. This suggests that the basic ethical questions to be asked are which values, whose, and why they are chosen.

The objective function of Cambridge Analytica's algorithms is simple and for sale: produce voting behavior aligned with the desires of its clients. China's social credit system and the broader policies and practices of its Cyberspace Administration offer an ostensibly more benign and values-promoting objective function – fostering social fairness and justice. Superficially, we seem to be confronted with two starkly different approaches to the algorithmic shaping of choices and conduct, and, more fundamentally, to two seemingly opposed systems of connectivity governance.

One is based on a choice-biased social mediation logic that maximizes both attention capture and experienced autonomy by ambiently reinforcing individuation-sustaining patterns of digital connectivity – a system epitomized by the American "market" approach to data governance that enables the fair and competitive pursuit of a putatively self-organizing and vibrantly "polyphonic" society. The other is based on a control-biased social engineering logic epitomized by China's "managerial" social credit system that maximizes cooperative potentials and relational integration – a system designed around centralized, population spanning manipulations of attentional focus and intentional dynamics in compositional pursuit of a stably "symphonic" society.

In contrast with Cambridge Analytica's crass instrumentalism, there is considerable appeal in China's stated commitment to developing a digital choice architecture for enhancing social welfare. If persons are presumed irreducibly relational if personal wellbeing is deemed a subordinate function of societal wellbeing, and if it is the state's responsibility to secure the conditions for societal wellbeing, then using algorithmic tools to shape citizen behavior and civic relations can be argued to be the state's moral responsibility. If a choice architecture *can* be designed to improve human–human, human–world, and human–technology–world relations then it *should* be designed and implemented. That, certainly, is the Chinese Communist Party's stance.

Of course, if the basic unit of ethical analysis is the individual and ideally autonomous human being, China's social credit system and its Cyberspace Administration's efforts to inculcate citizen and consumer conduct deemed desirable by the Chinese government amounts to coercive social engineering – a blatant violation of informational and communicative rights. Given that the same is true of Cambridge Analytica's electoral manipulations, it might be thought that the best and obvious alternative is to ensure that user preferences – and only user preferences – shape the choice architecture that emerges as their values and decisions are recursively integrated into the bespoke computations involved in curating their connective experiences.

This is the basic design of the search and recommendation algorithms currently employed, for example, by Google and Amazon. By taking attention to capture and the satisfaction of individual wants and desires as primary dimensions of the objective function of their machine learning systems, a recursively improving choice architecture emerges that is geared to ever more accurately anticipating and giving people what they want – "ethically laudable" systems for enhancing personal freedoms of choice.

Freedom of choice is undoubtedly preferable to its absence. But freedom of choice alone is a deficient determinant of freedom in the Buddhist sense of realizing virtuosic relational dynamics. Karma involves cyclic patterns of entanglement, with the experiential outcomes of enacted values and intentions serving also as volitional opportunities. The relational risk of self-improving search and recommendation algorithms is that users will be held hostage by their own past behaviors as their personalized choice architecture becomes so effective that they will always find "exactly" what they are looking for – news, entertainment, products, services, and social connection. The result over time will be the progressive loss of experiential and relational wilderness and "happy" residence on karmic "cul-de-sacs," leading lives in which it is never necessary to learn from mistakes or engage in adaptive conduct – lives in which course correction never seems either necessary or desirable.

This is troubling enough. But granted the relational nature of all things, the technological risks go much deeper. Among the causes of conflict, trouble, and suffering, Buddhism identifies as central the belief that each of us exists independently and possesses a unitary and abiding "self" that can exist independently of the body. Buddhist relational (rather than reductive) nondualism entails seeing all things as significantly interdependent – seeing how each thing *is* what it *means* for all others. Mind and body, the phenomenal and the physical, are constitutive *implications* of each other. Consciousness consists in the coherent differentiation of sensed and sensing presences, or the coherent differentiation of matter and what matters. That is, there is no 'hard problem' of explaining how the phenomenal arises out of the physical or how human motives arise out of neuro-molecular motions.

The brain is not the cause of consciousness. Rather, the relations among brains, bodies, and the environments with and within which they have coevolved constitute the *infrastructure* of consciousness. They are the result of what consciousness does – coherently elaborating differences – in much the same way that transportation infrastructure is a result of past transportation practices that then recursively shape further transportation practices. The ethical importance of this is that the infrastructure of human consciousness is both intracranial and extracranial, both personal and interpersonal and that it extends beyond our bodies into our natural environments, but also our social, cultural, and technological environments – including that of digital connectivity.

Field experimentation carried out via the computational infrastructure of digital connectivity to affect human choices, emotions, and conduct is analogous to inserting electrodes into the neural infrastructure of the brain for the purpose of producing certain phenomenal experiences or bodily actions. What Cambridge Analytica, Facebook, and the Chinese Cyberspace Administration are experimenting with is the insertion of algorithmic "electrodes" into the connective tissue of the socially embodied and enacted infrastructure of shared human consciousness – a process that is just as invasive and ethically fraught as inserting electrodes into the brain. Mass experiments in digitally attracting and exploiting human attention to shape emotions, beliefs, decision-making, and socialization are not and could never be value neutral.

If competing human values are infused into intelligent technology, it will scale up and deepen those conflicts. The ethical improvisation and relational virtuosity that will be required to resolve those conflicts will depend, according to Buddhism, on engaging in practices that will secure our most basic human right – the right to freedom-of-attention. Without freedom-of-attention there can be no freedom-of-intention, and without freedom-of-intention, we will remain bound by past karmic entanglements and thus unable to transform who we *are* by transforming what we *mean* to and for one another. We would be unable, in other words, to engage in the most important of all human arts – the ethical art of responsible course correction.

Virtue Ethics

By John Hacker-Wright

Virtue ethics can help us to understand the importance of transparency in algorithms that are designed to keep our attention focused on content and alter our desires and beliefs. Aristotle's views on character, in particular, can help us to understand what an authentic choice might be and thereby help us to agree on norms that promote such choice, as well as bring into focus what virtues users need to engage with social media technologies in a way that is not self-undermining.

In the Book VII of the *Nicomachean Ethics,* Aristotle outlines a useful typology of character. Most human beings fall in one of the following conditions: virtue, continence or strength of will, akrasia or weakness of will, and vice. Virtue is obviously a good state of character in which we aim at the right goals and we lack recalcitrant desires that pull us toward things we deem bad. It is an enviable condition since a fully virtuous agent would not have to struggle against her own desires to do what she judges best to do. The same is not true of a continent agent. She will have some desires that go against what she deems best to do, but it is characteristic of her to be reliably able to overcome those desires and to act on her judgment.

Both the virtuous agent and the continent agent have an intact capacity for choice, on Aristotle's view, because their overall desires will, in the end, come into line with their reasoning about what to do, even if that occurs with some struggle in the case of the continent agent. The incontinent agent, on the other hand, loses out to desires that go contrary to what they judge to be best. Incontinent agents find themselves doing things against their better judgment; they are likely to feel shame as a result. The reasoning and judgment that is guided by their correct conception of what to do is inert or impotent as they are ultimately guided by desires that they do not affirm.

Finally, there is the vicious agent. Aristotle gives two distinct pictures of the vicious agent. On one view, the vicious agent's desires are lined up behind their false idea of what is to be done, so what is chiefly wrong with the vicious agent is their false idea about what is good. In another view he gives, the vicious agent's desires are so disordered that vicious agents are, from a psychological standpoint, a wreck. Aristotle's view may be that while in principle there is a certain harmony in the vicious soul, they end up a wreck because of the nature of the thing they desire, which brings about ruin in the vicious agent.

If the vicious agent is necessarily a psychological wreck, then we have a very clear basis for recommending virtue; it should be pursued because anyone who does not, will end up in an unpleasant, tortured psychological condition. But it may be that people can be wholly committed to an end that seems entirely without redeeming value. So, we might imagine there could be someone who spends every waking hour of every day on social media or playing video games, and who feels no compunction or inclination to do anything else. In this case, we might want to say that there is a vice since the imagined individual is foolishly spending their life on something that is empty.

This is obviously a substantive moral judgment that the putatively vicious agent would reject. We might imagine that someone in this condition would be disturbed and feel wronged by non-transparent attempts to manipulate them into a different set of beliefs and desires. It is one thing if we acknowledge and deliberately choose to enlist assistance in changing behavior, say, by using time-tracking apps or apps that lock out social media access, but it is another if we are subject to attempts to alter our behavior without our knowledge. The upshot is that both virtuous and vicious agents may agree in rejecting efforts to manipulate their beliefs and desires without their knowledge.

Yet, that is what many social media and gaming technologies are deliberately aiming to do, with the result that many people who demonstrate the strength of will in other domains exhibit weakness of will in the face of these designs. Many of us were unaware, at least initially, of this deliberate effort to undermine our self-control on the part of designers. Possibly no development of human virtue will enable us to retain self-control in the face of such deliberate and sophisticated efforts. Hence, Vallor holds that an essential part of the solution is to demand tools that do not debilitate us.[87] That seems to me like a minimal demand. Ideally, we want tools that enhance rather than efface our agency.

Getting to the point of having such tools will surely require us all to deepen our understanding of psychology and technology to the extent necessary to understand how we are manipulated by technology. This knowledge seems to me to be part of what constitutes for us the general theoretical understanding of the human good and is a part of practical wisdom. Aristotle claims that in order to act well we need to have a general understanding of the human good. The requirement is not that we each need to have the level of expertise held by a doctor, psychologist, nutritionist, and physical trainer. That is specialized knowledge of the human good that is needed in certain circumstances. The sort of knowledge that is needed by any human being is more general: knowing what the good condition of a human being is so that I can consult a doctor or psychologist if I recognize that I am falling short of that condition, for instance.

I am claiming here, going beyond Aristotle, that the content of the general understanding of human good that is necessary for us may change over time, in response to new, widely shared circumstances that we face. It is plausible to think that we need a basic awareness of technology to function well in contemporary society, including awareness of how technology exploits our psychological vulnerabilities. This knowledge would give us some defense against techniques of mental malware that will no doubt persist even if we succeed in legislating better regulations requiring transparency in the design of algorithms. Indeed, sharing a general understanding of the need for such legislation will require the sort of understanding of the human good that I am outlining here. We need to have a basic understanding of what an unmanipulated, authentic human decision looks like in order to demand legislation that defends the possibility of such choice against technologies designed to undermine it.

Deontological Ethics

By Colin Marshall

As the Mental Malware case describes, Facebook (and other social media platforms) both influence and facilitate others influencing large numbers of people. Which ways of influencing other people are ethically permissible?

A deontological approach suggests two necessary conditions for ethical permissibility. First, any attempt to influence others must be constrained by respect, that is, by serious consideration of what their personal aims and projects are. Second, though all communication must involve some non-rational elements, deontology requires that influence prioritize honest communication that goes through our rational capacities. Neither of these conditions appears to have been met by key actors in the Mental Malware case. We can consider each condition in turn.

The first deontological condition on permissible influence can also be phrased as the requirement not to treat others *merely as means*. In other words, when we pursue some goal, we must avoid ways of achieving it that trample over other people's needs and projects, treating them as mere tools for getting what we want. This condition can explain why it is morally impermissible to make money off of addictive products that derail people's lives. The leaders of drug cartels, for example, achieve their financial aims with products that they know can destroy people's ability to live the lives they want to lead.

In the Mental Malware case, both Facebook and other entities that employed the platform (such as Cambridge Analytica and the Cornell researchers) seemed to treat Facebook's users merely as means. There is nothing inherently problematic about Facebook offering users a mechanism to maintain and build friendships, offering them opportunities for entertainment (such as personality quizzes), or collecting data about users' behavior. But things do become ethically problematic, according to the first deontological condition, when the platform attempts to draw users in ways that can negatively impact their lives, and when the platform facilitates the collection and use of data solely in order to generate profit and produce political outcomes that (in at least some cases) were clearly not in users' interests.

The second deontological condition on permissible influence is that it prioritizes honesty and appeals to others' rational capacities. Flagrant violations of this condition would include lies, brainwashing, and the use of alcohol to 'soften people up'. Deontologists have always recognized, though, that all interactions between people involve what Thaler and Sunstein call 'choice architecture', which appeals to what Kahneman calls 'System 1'. Clearly, when giving options, one option must be given first, and that ordering can impact people's decisions. Yet even if influence must *start* by an appeal to System 1, the deontological condition demands that it still *go through* System 2 (conscious, attentive reasoning), giving people at least the option of rationally assessing, and potentially rejecting the attempted influence.

There seem to have been various violations of this condition in the Mental Malware case. Voters who were presented with targeted ads typically did not know that the ads were based on their personal data. Voters in Honduras and other countries did not know that a significant number of "likes" around certain posts were inauthentic, or that other 'likes' were achieved via the Mathew Effect. Manipulative, inauthentic activity that *hinges* on secrecy and deception, not giving users the option of rationally assessing the attempted influence, fails to meet the second deontological condition for permissibility.

What could be done to improve things going forward? Increased transparency could, at least in principle, bring Facebook and entities using its platform closer to compliance with the two deontological conditions. After all, transparency is a form of honesty and can allow people to rationally assess whether they are willing to accept the impact of certain activities on their personal needs and projects. For example, Facebook users who know that Facebook tracks their activity and personalizes the ads they see can make a rational assessment of whether to continue using Facebook (in light of their own personal needs and projects) and decide whether to click on those ads.

There are at least two significant limits to how much improvement can come from transparency in this context. First, for transparency to be morally significant, the information presented must be *intelligible* – it is not enough to present people with long user agreements filled with legal jargon few can understand, nor is it enough to make public key algorithms that most users lack the technical abilities to process. Second, some manipulative techniques can continue to undermine people's rational processes even when we are made aware of them. Given the deep social structure of human psychology, the lure of well-crafted social rewards (such as 'likes' and followers) is extremely strong and can draw us into activities that go against our needs and projects, even when we are aware of what is happening. For transparency to be ethically significant by deontological standards, then, it must be tailored to users in a way that facilitates meaningful, informed choices based on what matters to them.

Notes

1. Elna Schultz, "The Sophisticated Tech Predicting If an Ad Will Work," *BBC News*, March 6, 2022, https://www.msn.com/en-us/money/personalfinance/the-sophisticated-tech-predicting-if-an-ad-will-work/ar-AAUHEaQ. Collinge designed the 'Ad of the Decade', the most popular advertisement in modern UK history. It was an advert for Cadbury Dairy Milk featuring an actor in a gorilla suit drumming along to the Phil Collins Song, "In the Air Tonight," You Tube, March 30, 2011, https://www.youtube.com/watch?v=yYYtgsoMap4.
2. Sam Meredith, "Facebook-Cambridge Analytica: A Timeline of the Data Hijacking Scandal," *CNBC News*, April 10, 2018, https://www.cnbc.com/2018/04/10/facebook-cambridge-analytica-a-timeline-of-the-data-hijacking-scandal.html.
3. Meredith, "Facebook-Cambridge Analytica."
4. Meredith, "Facebook-Cambridge Analytica."
5. Meredith, "Facebook-Cambridge Analytica."
6. Stephen Soldz, and George E. Vaillant, "The Big Five Personality Traits and the Life Course: A 45-Year Longitudinal Study," *Journal of Research in Personality*, 33, no. 2 (1999): 208–232, doi: https://doi.org/10.1006/jrpe.1999.2243. The 'Big 5' personality traits used in this model are extroversion/introversion, openness, conscientiousness, agreeableness, and neuroticism.
7. Meredith, "Facebook-Cambridge Analytica."
8. Meredith, "Facebook-Cambridge Analytica."
9. Christopher Wylie, *Mindf*ck: Cambridge Analytica & the Plot to Break America* (New York: Random House, 2019), Chapter 7, Loc. 3/238.

10. Wylie, *Mindf*ck*, Chapter 6, Loc. 132/185.
11. Carole Cadwalladr, and Emma Graham-Harrison, "Revealed: 50 Million Facebook Profiles Harvested for Cambridge Analytica in Major Data Breach," *The Guardian*, March 17, 2018, https://www.theguardian.com/news/2018/mar/17/cambridge-analytica-facebook-influence-us-election.
12. Wylie, *Mindf*ck*, Chapter 6, Loc. 180/185.
13. Roger McNamee, *Zucked: Waking up to the Facebook Catastrophe* (New York: Penguin Books, 2020), Chapter 10, Loc. 77/204.
14. Cadwalladr, "Revealed: 50 Million Facebook Profiles Harvested.".
15. Cadwalladr, "Revealed: 50 Million Facebook Profiles Harvested."
16. Cadwalladr, "Revealed: 50 Million Facebook Profiles Harvested."
17. Wylie, *Mindf*ck*, Chapter 6, Loc. 7/185.
18. Linda Killian, *The Swing Vote: The Untapped Power of Independents* (New York, St. Martin's Press, 2012).
19. Schultz, "The Sophisticated Tech."
20. McNamee, *Zucked*, Chapter 13, Loc. 88/350.
21. McNamee, *Zucked*, Chapter 1, Loc. 39/175.
22. McNamee, *Zucked*, Chapter 11, Loc. 38/129.
23. Wylie, *Mindf*ck*, Chapter 12, Loc. 55/368.
24. Wylie, *Mindf*ck*, Chapter 12, Loc. 55/368.
25. Christopher Wylie, *Mindf*ck*, Chapter 12, Loc. 337/368 [emphasis in original].
26. William J. Brady, M.J. Crickett, and Jay J. Van Bavel, "The MAD Model of Moral Contagion: The Role of Motivation, Attention, and Design in the Spread of Moralized Content Online," *Perspectives on Psychological Science*, 15, no. 4 (2020): 978–1010, doi: https://doi.org/10.1177/1745691620917336, 978.
27. Brady, "The MAD Model of Moral Contagion," 979. Brady *et al.* define moralized content as that which evokes "ideas, objects or events typically construed in terms of the interests or good of a unit larger than the individual."
28. Björn Lindstrom, Martin Bellander, David T. Schultner, Allen Chang, Philippe N. Tobler, and David M. Amodio, "A Computational Reward Learning Account of Social Media Engagement," *Nature Communications,* 12 (2021): Article No. 1311, DOI: doi.org/10.1038/s41467-020-19607-x, 2.
29. Lindstrom, "Computational Reward Learning," 3.
30. Lindstrom, "Computational Reward Learning," 7.
31. Adam D.I. Kramer, Jamie E. Guillory, and Jeffrey T. Hancock, "Experimental Evidence of Massive-Scale Emotional Contagion Through Social Networks," *Proceedings of the National Academy of Sciences of the United States,* 111, no. 24 (2014): 8788–8790, doi: https://doi-org.proxy.lib.sfu.ca/10.1073/pnas.1320040111.
32. Catherine Flick, "Informed Consent and the Facebook Emotional Manipulation Study," *Research Ethics*, 12, no. 1 (2016), 16.
33. Flick, "Informed Consent and the Facebook Emotional Manipulation Study," 16.
34. Kramer, "Experimental Evidence of Massive-Scale Emotional Contagion."
35. Flick, "Informed Consent and the Facebook Emotional Manipulation Study," 16.
36. Flick, "Informed Consent and the Facebook Emotional Manipulation Study," 16. It is also notable that the journal that published the study, the *Proceedings of the National Academy of Sciences*, required institutional ethics board approval for all studies involving human subjects.
37. Flick, "Informed Consent and the Facebook Emotional Manipulation Study," 19.
38. Flick, "Informed Consent and the Facebook Emotional Manipulation Study," 14.
39. Flick, "Informed Consent and the Facebook Emotional Manipulation Study," 23.
40. Flick, "Informed Consent and the Facebook Emotional Manipulation Study," 24.
41. Flick, "Informed Consent and the Facebook Emotional Manipulation Study," 26.
42. Wylie, *Mindf*ck*, Chapter 7, Loc. 46/238.

43. Wylie, *Mindf*ck*, Chapter 8, Loc. 157/240.
44. Richard H. Thaler, and Cass R. Sunstein, *Nudge: The Final Edition* (New York: Penguin Books, 2021), Introduction Loc. 25/176.
45. Thaler, *Nudge*, Introduction, Loc. 30/176.
46. Thaler, *Nudge*, Introduction, Loc. 112/176.
47. Thaler, *Nudge*, Chapter 1, Loc. 197/208.
48. Svoboda, "The 'Neuropolitics' Consultants."
49. Daniel Kahneman, *Thinking: Fast and Slow* (Toronto: Anchor Canada, 2013).
50. Schultz, "The Sophisticated Tech." Collinge designed the 'Ad of the Decade' – the most popular advertisement in modern UK history. It was an advert for Cadbury Dairy Milk featuring an actor in a gorilla suit drumming along to the Phil Collins Song, "In the Air Tonight," *You Tube*, March 30, 2011, https://www.youtube.com/watch?v=yYYtgsoMap4.
51. Julie Carrie Wong, and Jeff Ernst, "Facebook Knew of Honduran President's Manipulation Campaign – and Let it Continue for 11 Months," *The Guardian*, April 13, 2021, https://www.theguardian.com/technology/2021/apr/13/facebook-honduras-juan-orlando-hernandez-fake-engagement?via=webuproar.
52. Wong, "How Facebook Let Fake Engagement Distort Global Politics."
53. Wong, "How Facebook Let Fake Engagement Distort Global Politics."
54. Wong, "How Facebook Let Fake Engagement Distort Global Politics."
55. Wong, "How Facebook Let Fake Engagement Distort Global Politics."
56. Craig Silverman, Ryan Mac, and Pranav Dixit, "'I Have Blood on My Hands': A Whistleblower Says Facebook Ignored Global Political Manipulation," *Buzzfeed News*, September 14, 2020, https://www.buzzfeednews.com/article/craigsilverman/facebook-ignore-political-manipulation-whistleblower-memo.
57. Wong, "How Facebook Let Fake Engagement Distort Global Politics."
58. Wong, "How Facebook Let Fake Engagement Distort Global Politics."
59. Meta, "Inauthentic Behavior," Facebook Community Standards, current as of July 1, 2022, https://transparency.fb.com/policies/community-standards/inauthentic-behavior/.
60. Daniel Rigney, *The Matthew Effect: How Advantage Begets Further Advantage* (New York: Columbia University Press, 2010), 11.
61. English Standard Verse Bible, Gospel According to Mathew 13:12, (2001), https://www.esv.org/Matthew+13/.
62. Rigney, *The Matthew Effect*, 9.
63. Rigney, *The Matthew Effect*, 9.
64. Rigney, *The Matthew Effect*, 16.
65. Rigney, *The Matthew Effect*, 83.
66. Wong, "How Facebook Let Fake Engagement Distort Global Politics."
67. Wong, "How Facebook Let Fake Engagement Distort Global Politics."
68. Wong, "How Facebook Let Fake Engagement Distort Global Politics."
69. Wong, "How Facebook Let Fake Engagement Distort Global Politics."
70. United States Department of State Report. "2020 Country Report on Human Rights Practices: Azerbaijan." United States Department of State, Bureau of Democracy, Human Rights and Labor. Report. March 31, 2020, https://www.state.gov/reports/2020-country-reports-on-human-rights-practices/azerbaijan/.
71. Wong, "How Facebook Let Fake Engagement Distort Global Politics."
72. Julia Carrie Wong, "Revealed: The Facebook Loophole that Lets World Leaders Deceive and Harass their Citizens," *The Guardian*, April 12, 2021, https://www.theguardian.com/technology/2021/apr/12/facebook-loophole-state-backed-manipulation?CMP=series_embed_box.
73. Wong, "Revealed: The Facebook Loophole."
74. Wong, "How Facebook Let Fake Engagement Distort Global Politics."

75. Silverman, "I Have Blood on My Hands."
76. Silverman, "I Have Blood on My Hands."
77. Wong, "Revealed: The Facebook Loophole."
78. Wong, "How Facebook Let Fake Engagement Distort Global Politics."
79. Wong, "Revealed: The Facebook Loophole."
80. Jennifer Conrad, and Will Knight "China Is About to Regulate AI – and the World is Watching," *Wired*, February 22, 2022, https://www.wired.com/story/china-regulate-ai-world-watching/.
81. Cyberspace Administration of China, "The State Internet Information Office and Other Four Departments Issued the 'Internet Information Service Algorithm Recommendation Management Regulations,'" *Cyberspace Administration of China: Laws*, January 4, 2022, trans. Google Translate, http://www.cac.gov.cn/2022-01/04/c_1642894606258238.htm. The regulations Came into force on March 1, 2022. See also: Cyberspace Administration of China, "Notice on Launching the Special Action of 'Qinglang 2022 Comprehensive Algorithm Governance," *Cyberspace Administration of China: Laws*, April 8, 2022, trans. Google Translate, http://www.cac.gov.cn/2022-04/08/c_1651028524542025.htm, for a description of some specific regulatory actions.
82. Cyberspace Administration of China, "The State Internet Information Office."
83. Jennifer Leaning, "War Crimes and Medical Science," *British Medical Journal*, 313 (1996): 1413, doi: https://doi.org/10.1136/bmj.313.7070.1413.
84. *U.S.A.* v. *Karl Brandt et al.*, Trials of War Criminals before the Nuremberg Military Tribunals, August 20, 1947 (the 'Doctors' Trial'), accessed July 1, 2022, http://nuremberg.law.harvard.edu/nmt_1_intro.
85. *U.S.A.* v. *Karl Brandt et al.*
86. United States Holocaust Memorial Museum, "The Nuremberg Code," *Holocaust Encyclopedia*, n.d., accessed July 1, 2022, https://encyclopedia.ushmm.org/content/en/article/the-nuremberg-code.
87. Shannon Vallor, "Moral Deskilling and Upskilling in a New Machine Age: Reflections on the Ambiguous Future of Character," *Philosophy & Technology*, 28, no. 1 (2015): 119, doi: https://doi.org/10.1007/s13347-014-0156-9.

9 AI and Nonhuman Animals

The day may come when the rest of the animal creation may acquire those rights which never could have been withholden from them but by the hand of tyranny … Is it the faculty of reason or perhaps the faculty of discourse? But a full grown horse or dog is beyond comparison a more rational, as well as a more conversable animal, than an infant of a day, or a week, or even a month, old. But suppose they were otherwise, what would it avail? The question is not, Can they *reason*? nor Can they *talk*? but, *Can they suffer?*

~ (Jeremy Bentham, 1789)[1]

By Peter Singer and Yip Fai Tse

Humans are not the only stakeholders that are impacted by artificial intelligence (AI) systems and data science. As we will show, nonhuman animals are impacted too. If nonhuman animals have morally relevant interests as, on a wide variety of ethical viewpoints, they do, we have reasons for concern about the impact of technologies on them.

Assessing the ethics of AI on nonhuman animals is fundamentally different from that of humans. First, nonhuman animals do not – and arguably cannot – *actively* and *knowingly* participate in the process of design, development, or deployment of AI systems, nor can they provide meaningful feedback on their use, at least at present.[2] Their involvement in AI systems is passive, without consent, and often without even a meaningful assessment of the impact of the system on their well-being.[3] Worse still, nonhuman animals usually have no way of complaining or objecting when AI systems have a negative effect on them – they don't understand what is harming them, let alone how to report it. The impact of AI systems on other living beings is therefore entirely within humans' hands.

Second, nonhuman animals do not enjoy the same protections – whether they be legal, cultural, or structural – as do humans. Some nonhuman animals have none. Indeed, nonhuman animals do not, as yet, have legal personhood and thus the legal protections that flow from this.[4] Those laws that do protect animal welfare are poorly enforced, and even egregious harms perpetrated against animals are rarely prosecuted.[5] For example, when a self-driving car

DOI: 10.1201/9781003293125-9

takes the life of a human, it will be scrutinized by courts, governments, and society. The same is less likely to be true if a nonhuman animal is killed by a self-driving car. Most people will not even think of it as an ethical problem, let alone a legal one. In some cases the plight of nonhuman animals is deliberately hidden from the public, for example by the "ag-gag" laws in several states in the United States that ban the use of undercover investigations against factory farms.[6] Our neglect of nonhuman animals, combined with their inability to meaningfully assert their interests within our political and legal systems – indeed even within the broader moral framework of our culture – means that the harms caused to nonhuman animals by AI are likely to remain hidden and unaddressed. They will only be discovered if thoughtful humans who care about the interests of nonhuman animals are willing and able to investigate just what it is that AI is doing to them.

Third, while there is only one human species left, there are a multitude of nonhuman animal species in the world. The differences between the various species are far wider than the differences between humans. This gap means that animals of different species have strikingly different interests, and therefore the impact that AI systems have on them will raise problems that differ greatly from one species to another. For example, this raises profound questions about the moral status of different species: Which species deserve moral consideration? Mammals? Birds? Reptiles? Amphibians? Crustaceans? Mollusks (which include the octopus)? Insects? The level of evidence scientists has about the sentience of these animals differs considerably, and hence the ethical treatment of each of these kinds of animals might also differ greatly.

Despite these difficulties, the impact of AI on animals other than ourselves merits serious attention. Here, we assess some of the impacts of AI and data science on nonhuman animals. Out of a wide range of possible cases, we have chosen two for discussion: how algorithmic bias can impact nonhuman animal welfare, and the use of AI in factory farming.

Algorithmic Bias

Algorithmic bias is an important topic in the ethics of AI and data science. Most discussions about it relate to gender and racial bias – to a lesser extent – biases related to age, religious beliefs, sexuality, social status, and educational background. But our research shows that algorithmic biases against nonhuman animals exist and can have a negative impact on their well-being. We have found algorithmic biases against nonhuman animals in search engines, recommendation algorithms, and artificial language models.

First, we introduce some necessary terminology. While the terms 'racism,' 'sexism,' and even 'ageism' are well-known and often used, the term 'speciesism,' introduced by Richard Ryder in 1970[7] is a counterpart to these terms and is used to describe biases and prejudices against other beings based on their species. This applies to situations where humans hold prejudicial attitudes that discriminate in favor of their own species, *Homo sapiens*, and against other species. But it also applies to situations where humans hold biased attitudes toward different species, for example, when we accept treating pigs in ways that we vehemently reject for dogs.

Recommendation algorithms are used in social media platforms, shopping apps, and planning apps, all of which affect the lives of animals. For example, food or restaurant recommendation apps and food planning apps can increase – or decrease – our consumption of meat, dairy, and eggs. Video recommendations can affect the number of animal cruelty videos people watch. Our research has shown that not all animal cruelty videos are taken down; videos of the torture of certain species, such as aquatic animals and rats, are abundant on YouTube.

Language Models and Databases

Studies have found that sexist,[8] racist,[9] and ageist[10] language affects people's attitudes. Speciesism is also present in human languages, including but not limited to English and Chinese and it affects our attitudes toward animals.[11] For example, Kunst and Hohle found that describing meat production as 'harvesting' rather than by more direct terms such as 'slaughtering' or 'killing' tended to reduce empathy. Referring to items on restaurant menus as a type of meat rather than a type of animal – 'beef' or 'pork' versus 'cow' or 'pig' – also reduced empathy and disgust and increased diners' willingness to consume animal products and decreased their willingness to consume vegetarian dishes.[12] Most machine learning models that process language will reproduce and reinforce these entrenched speciesist patterns of language, which we argue lowers empathy toward nonhuman animals and further promotes their consumption.

WordNet® is a large lexical database of different languages including English, often used in natural language processing. It contains speech patterns showing speciesist biases against some animals. For example, the word "chicken" was given the following senses: "<noun.food> (the flesh of a chicken used for food)", or "(a domestic fowl bred for flesh or eggs;)". Both descriptions – indeed all five descriptions we examined – are speciesist and demeaning to chickens, as they focus on the value of the animal as food rather than as a living being in possession of an intrinsic worth. It is also worth mentioning that the sense in which the word "chicken" means flesh comes before the second meaning in which "chicken" means the animal that is raised to become food. This might influence language models based on the database, as the rankings of the senses have statistical and structural implications.[13] Indeed, speciesist language is also used to demean human beings, as when the word 'dog' is used to mean a 'dull and unattractive girl or woman,' or when 'fox' is used to mean a "shifty deceptive person,' or when a 'chicken' refers to a 'person who lacks confidence, is irresolute and wishy-washy'.

The speciesist language we see reflected in these databases may well reflect actual speciesist attitudes in the language we use, rather than mere artifacts of the language databases themselves.

Gallus domesticus, for example, is the scientific name for the chicken, which reflects their history as a species specifically bred for human use and consumption.[14] Here, we introduce an important consideration: even where human attitudes change and our language comes to reflect updated usages and moral development, language databases used to train machine learning systems may not be. This is the case for WordNet, which has not been updated since the 1990s.[15] WordNet captures the attitudes, including toward nonhuman animals, of that age. Despite containing some outdated language, WordNet continues to exert its influence on human attitudes through its use in natural language processing applications widely used in businesses, education, and the media. A concerted effort to remove speciesist biases from language databases is needed to keep these biases from solidifying and being propagated in real-world applications.

Speciesist biases influence the language models trained on datasets that contain them. One of the models investigated was Word2Vec, which can be trained using text corpora like Google News, Wikipedia, or Twitter tweets. We selected three groups of words: five words that typically describe humans (human, woman, man, child, person); five words describing non-farmed animals that people typically think of when they think 'animal' (dog, cat, rabbit, parrot, horse); and five words describing animals who are routinely farmed (cow, pig, chicken, sheep, hen), and calculated the mean word similarities between the three groups and a list of ten words related to moral status (sentient, conscious, intelligent, right, worthy, esteem, interest, protection, protected, deserve). All trained models reveal speciesist tendencies (Figure 9.1). For example, humans are far more closely associated with words related to moral status than nonhuman animals, and nonhuman animals typically thought of as companion animals are more closely associated with these words than farmed animals.

There are efforts to "debias" language models and natural language processing techniques for racist and sexist biases.[16] But to our knowledge, there is not yet any such effort regarding speciesism.

Bias in Search Engine Algorithms

It is widely accepted now that search engines give biased results in relation to gender and race and that these biases have real-world impacts.[17] What is just as significant, but much less often discussed, are search engine biases concerning political opinion,[18] age,[19] and geography.[20] Our research supports the hypothesis that speciesist biases are present as well. We argue that this will propagate speciesist attitudes and affect our attitudes toward animals in ways that are similar to how sexist, racist, ageist, and elitist search results can affect attitudes toward other humans.

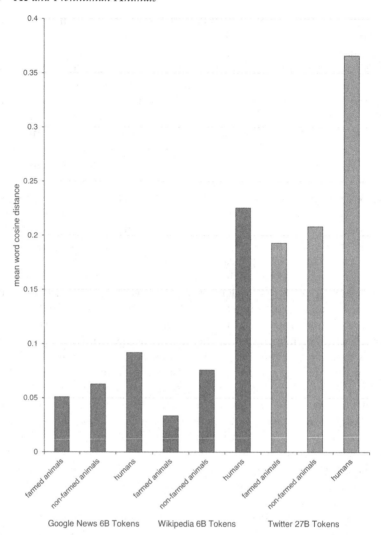

Figure 9.1 Word2Vec natural language model trained using Google News, Wikipedia, and Twitter. We tested three groups of words: five words that typically describe humans (human, woman, man, child, person); five words describing non-farmed animals that people typically think of when they think 'animal' (dog, cat, rabbit, parrot, horse); and five words describing animals who are routinely farmed (cow, pig, chicken, sheep, hen). We calculated the mean word similarities between the three groups and a list of ten words related to moral status (sentient, conscious, intelligent, right, worthy, esteem, interest, protection, protected, deserve). Each trained model reveals speciesist tendencies. We excluded invertebrate farmed animals here in order to have a better matched comparison, and not because they are of lesser value. If invertebrates are included, the results are even more significant. Fish are excluded because they are both commonly farmed animals and common companion animals and cannot be appropriately compared in the present model.

Table 9.1 Presence of classes of different animals in top 100 results in Google image search of certain keywords

Searched words	Top result	Dominant result in top 100	Terrestrial farmed animals	Aquatic animals	Terrestrial wild animals	Animals used in experiments
"animal cruelty"	dog	dog (88)	4	0	0	6
"animal abuse"	dog	dog (87)	6	0	2	2
"animal charity"	no animal	dog (93)	2	0	1	6
"animal protection"	no animal	dog (50)	7	1	11	5

We conducted a search results analysis for Google image search (as it is much less affected by IP addresses than the regular search).[21] As of 4th April 2022, using the private mode in the Edge browser,[22] and refraining from clicking anything after the search (to not look at the branched search results), we ran searches and collected some data. Here, a 'depiction' includes both cartoon representations as well as photos. The results are given in Table 9.1.

The search results returned are not representative of the treatment that nonhuman animals actually face. For instance, factory-farmed animals suffer much more than dogs in terms of both the extent and scale of cruelty. This underrepresentation of farmed animals compared to other nonhuman animals is probably not surprising, as it obscures what we come to understand as animal cruelty. Just consider leg lameness in broiler chickens. A study found that in the UK broiler chicken raising industry, "At a mean age of 40 days, over 27.6% of birds in our study showed poor locomotion and 3.3% were almost unable to walk."[23] These situations are clearly painful for the chickens and would be considered animal cruelty were it to happen to dogs or cats. In terms of numbers, according to FAOSTAT, the U.K. slaughtered 1.15 billion broiler chickens in 2020. Even if we only take the cases in which the chickens can barely walk to be cruelty (which in our opinion, is too high a bar for what counts as animal cruelty), the number of chickens involved was 38 million. The number of domestic dogs in the United Kingdom was 9.6 million in 2020,[24] and the number of reports to RSPCA of cruelty toward dogs was 56,563 (out of 1,016,455 reports) in 2020.[25] The disproportionate representation is even more severe when it comes to aquatic animals. According to fishcount.org.uk, the estimated number of wild-caught finfish[26] is somewhere between 1 and 3 *trillion* each year, and the estimated range of the number of farmed finfish slaughtered is about 51–167 billion each year. Many, if not most, of these finfish, endure extremely painful experiences when they are trawled by nets, raised in crowded enclosures, transported, and slaughtered.[27] And let's not forget about the invertebrate animals (such as crustaceans and cephalopods), for which scientists[28] and governments[29] are starting to identify more evidence that supports the idea that they are sentient. They are often omitted even by animal advocates, and the number of individuals is not well

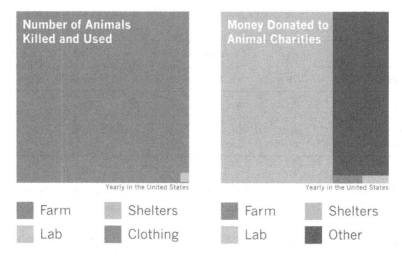

Figure 9.2 Comparison of money donated to animal charities with the number of animals used and killed in different settings. Courtesy of Animal Charity Evaluator.

researched and documented but could be even higher than finfish. Despite their dominance in numbers, all aquatic animals are close to being not represented in search engines, social media, and encyclopedias when the search is generalized to the level of "animal".

There are fewer charities to assist farmed, wild, and experimented-upon animals. We see that domesticated companion animals – especially dogs – tend to dominate the search engine rankings for those who search to donate their money to animal welfare (Table 9.1). Charities that assist with the welfare of farmed, wild, and invertebrate animals receive lower exposure, and therefore fewer donations (Figure 9.2). This is itself an ethical problem that can be attributed to the algorithms that determine search engine rankings. We found the above biases not only in Google's search engine rankings, but also in Bing, Duckduckgo, Yandex, and in Chinese-language searches on Baidu.[30] Speciesist biases seem to be everywhere.

For search results in Google image search, our research shows that even a slight change of wording drastically affects the search results of similar concepts related to nonhuman animals. For example, "farm animal" and "farmed animal" may seem like very similar concepts, but they reflect different attitudes toward the animals. The organization Sentient Media explains that the "term farmed animal can be distinguished from farm animals since the latter suggests that certain species of animals typically chosen for farming are biologically designed for human consumption."[31] Sentient Media and some other nonhuman animal advocates reject this speciesist suggestion. Additionally, the word "farm" seems to have a generally good connotation. When "farm animal" was searched, the top result and most (76) of the top 100 results are cartoon animals in green, cozy, open pastures or barns, with

Table 9.2 Presence of classes of different situations animals in top 100 results of Google image search of certain keywords

Searched words	Top result	Cartoon animals	Real animals depicted in desirable 'free-range' conditions	Real animals depicted in crowded conditions	Photo/cartoon depiction of humans causing apparent distress to animals
"farm animal"	Cartoon animals	76	22	0	1
"farmed animal"	Cartoon animals	6	37	31	2
"animal farm"	All search results returned depictions of George Orwell's 1945 *Animal Farm*.				
"factory farm"	typical factory farms	0	0	97	1
"animal agriculture"	typical factory farms	8	41	40	0

some of the animals with "smiling faces" (Table 9.2). When "farmed animals" was searched, although the top result was still a cartoon, only 6 out of the top 100 results were cartoon animals. A similar pattern was found while comparing the search results of "factory farm" and "animal agriculture". The more neutral sounding words "animal agriculture" did return more neutral results. The search results of "farm animal" and "animal agriculture" are not representative of the true percentage of farmed animals that live in crowded situations. According to the Sentience Institute, 99% of terrestrial farmed animals live in "factory farms", a term they, and a lot of farmed animal advocates, use to describe "Concentrated Animal Feeding Operations",[32] a term used by the U.S. Department of Agriculture and U.S. Environmental Protection Agency. The American Society for the Prevention of Cruelty to Animals, on the other hand, estimates that "95% of farm animals in the U.S. are raised in factory farms."[33]

AI in Factory Farming

If speciesist algorithmic bias affects nonhuman animals indirectly by affecting the attitude of humans toward them, some other AI systems and data science applications affect the lives of nonhuman animals directly. The use of AI and data science in factory farming is an example.

With biometric sensors and environmental sensors, a lot of data can be gathered from farmed animals and a variety of pattern recognition and prediction can be done. These allow new profitable opportunities, such as cost reduction, risk reduction (less chance of high or 100% mortality events), or higher perceived quality food products.

In recent years, such data-centric systems have begun to be tested or applied in the factory farming industry. These systems can collect data from farmed animals such as their body temperature, sounds, weight, growth rate, and visible health problems such as injuries, bruises, and parasites. Machine learning models are created to relate physical parameters to important profitability metrics, such as disease, mortality, growth rate, eating patterns, etc. These systems can then put forward recommendations, or sometimes even direct actions through robotics, such as varying the quantity of food provided, disease treatments, or culling.[34]

In the raising of fish in enclosed systems, two major sources of risks and costs are diseases and under or overfeeding. The former affects mortality rate and product quality, while the latter affects either the growth rate (underfeeding) or cost of feed spent, water quality, and pollution (overfeeding). The traditional model for tackling these two problems is experience, intuition, and luck. Even though some education and knowledge-spreading mechanisms exist within the industry, each practitioner can still be said to have their own unique models.

But a new mode of working in the industry has emerged in recent years. Some aquaculture service companies like Aquabyte work with aquaculture companies to let them use Aquabyte's gadgets, application programming interface, specialized computers, and most importantly, machine learning models that do all the important learning, modeling, and decision-making. For example, sea lice, an extradermal parasite, is often seen on the skins of salmon and is a common problem for the salmon-raising industry globally. Due to the risk of sea lice from farmed salmon spreading to wild salmon and other aquatic animals, various jurisdictions require close monitoring and control of sea lice levels for salmon-raising facilities.[35] The traditional method for monitoring sea lice is to pick up a sample of the salmon and literally count the number of sea lice visually.[36] Aquabyte uses computer image recognition techniques to count the number of sea lice on each salmon from real-time images taken underwater,[37] which not only reduced labor but also there is no need to take salmon out of the water – an action that risks damaging their health. The image recognition algorithm improves over time, using human-verified data to expand the training dataset the algorithm is based on. Besides sea lice, the problem of under- or overfeeding can also be tackled using data and machine learning. Umitron, a competitor of Aquabyte, uses machine learning algorithms to analyze video data to calculate fish appetite, in terms of an index. The algorithm "observes fish-eating behavior along with the circumstances, before scoring their appetite and presenting the data in an easy-to-understand chart." They claim that "this allows farmers to make data-driven decisions when feeding their fish."[38]

Machine learning can also be used to train algorithms to 'identify' the welfare states of nonhuman animals – both companion animals and factory-farmed animals.[39] According to our research – although common sense would not be far behind – there is a common problem with algorithms that purport to identify welfare/behavioral states in nonhuman animals. All the data that goes into training the algorithms is labeled by humans and thus burdened by *our own* values and empirical judgments rather than accurately reflecting the animal's own interests and subjective determinations of what the *nonhuman animal itself values*. The accuracy and trustworthiness of the algorithms are thus limited to how trustworthy and accurate we humans have been in decoding the subjective lives of nonhuman animals in the first place.[40] Further research into how nonhuman animals can label and promote their own welfare using AI systems is urgently needed; but to avoid bias, it needs to be done by people who are not stakeholders in the factory farming industry.

Commentary

Jewish Ethics

By Daniel Sinclair

The two fundamental biblical principles governing the traditional Jewish approach to animals are human domination over all other living creatures enshrined in God's blessing to Adam and Eve: "...and have dominion over the fish of the sea, and over the fowl of the heaven, and over every animal that swarms upon the earth";[41] and the prohibition on causing suffering to any living creature in the course of exercising that dominion.

Maimonides identifies the angel's rebuke to Balaam for angrily striking his she-ass as the source for the prohibition.[42] Other authorities locate its source in various laws such as the obligation to help unload an animal that has collapsed under its burden[43] and the prohibition on muzzling an ox when it is threshing.[44] It is significant that under biblical law, animals are also required to rest on the Sabbath[45] and must be given access to pasture so that they are not only relieved of labor but also provided with satisfaction and contentment.[46] Moreover, the Talmud rules that in certain cases, the preservation of animal welfare overrides the prohibitions of the Holy Day.[47]

The domination principle justifies using animals for a wide range of human needs including food. In the beginning, however, humans were vegetarians,[48] and permission to consume the flesh of animals was only given in the postdiluvian era with the proviso that the animal was to be slaughtered prior to its consumption.[49] According to one view, the reason for this concession was that humankind degenerated physically after the Flood, and as a result, people were no longer able to satisfy their nutritional needs unless they ate meat.[50] Another explanation for the change given by R. Kook (1865–1935) in his monograph – "The Vision of Vegetarianism and Peace" – is the extremely low level of ethics and spirituality in the postdiluvian period which manifested itself in a lust for meat. In order to prevent cannibalism, humans were permitted to eat the flesh of animals but not of their own kind.[51] R. Kook looks forward to an age in which humankind will evolve ethically to the extent that it will totally reject the notion that it is permissible to consume the flesh of any living creature. In the interim, however, animals may be slaughtered for their meat but as Maimonides points out,[52] the killing must be done as painlessly as possible; hence, the plethora of regulations governing ritual slaughter in Jewish law, the object of which is to ensure a minimum of animal suffering in the process.

The avoidance of cruelty is not only the motivation for the laws of ritual slaughter; it also functions as a meta-legal ethical restriction on the use of animals for meat. The Talmud recounts that R. Judah the Prince, the redactor of the Mishnah, was giving a public lecture when a calf which was being led to the slaughter entered the hall and thrust its head under the corner of his robe lowing piteously. R. Judah told the calf: "Go! You were created for this purpose [to be slaughtered]." At that moment it was said in Heaven: "since he failed to show mercy, let him experience suffering," and he was afflicted with a painful ailment for thirteen years. One day, his maid was sweeping the house when she came upon a litter of newly born weasels and was about to sweep them up when R. Judah said: "Let them be; as it is written: 'And His mercies extend to all His creations.'"[53] They said in Heaven: "'since he shows mercy, let us show mercy to him' and his pain ceased."[54] According to the Talmudic commentators, R. Judah was punished because in his role as the leader of his generation, his responsibility was to instruct his listeners in the meta-legal principle of cruelty prevention by manifesting empathy and saving the calf; in the specific circumstances of the case, relying upon the letter of the law was unacceptable.

Contemporary animal farming has given rise to a host of ethical challenges well beyond those of painless slaughter and the traditional uses of livestock. Technology, including AI, effectively controls the lives of animals from birth to death and is particularly adept at concealing the cruelty which is often involved in modern animal husbandry.[55] One ameliorative Jewish initiative in this area is the Conservative Movement's 2011 *Magen-Tzedek* project which seeks to label food "ethically obtained" in addition to the more well-known Kosher label (*hekhsher*) attesting to its ritual suitability.

Deontological Ethics

By Colin Marshall

As the "Animals and Nonhuman AI" case describes, AI and related technologies have a massive impact on nonhuman animals (hereafter: "animals") and our attitudes toward them, at a scale that was unimaginable a few decades ago.

All versions of deontology agree that we have some moral obligations relating to animals. However, some versions of deontology hold that animals are morally important in the same way humans are, while other versions hold that their moral importance is of a different, lesser kind. We can consider this case from each perspective in turn.

If animals are morally important in the same way that humans are, then deontology implies that they also deserve moral respect. For deontologists, one key component of moral respect is taking others' needs and projects seriously, and not trampling over them in pursuit of our own goals (i.e., using them as mere means). What this amounts to in particular cases depends on what needs and projects are involved: laughing at a comedian's slapstick antics can be perfectly respectful, but laughing at another person's physical injuries can be deeply disrespectful – failing to take their physical needs seriously, and using them as a mere means for entertainment.

Many forms of bias involve a failure of respect. Systematic racial bias, for example, can involve members of one race failing to take the needs and projects of other races seriously, making it easier for the former to use the latter as mere means. Algorithmic biases in search engines can perpetuate and amplify such failures of respect. When searches for "animal abuse" and "animal protection" misleadingly show many more results related to dogs than to other species, this is itself a failure to take other species' needs seriously and can perpetuate existing failures of respect in people who use the search engine. Such failures make it easier for people to trample over other animals (sometimes literally) in pursuit of their own, often superficial, goals, such as culinary pleasures. Similarly, efficiency-increasing technologies used in factory farming can involve and perpetuate failures of respect, treating animals as mere data points in a profit-generating process, instead of as beings whose needs themselves call for consideration.

But what if we deny that animals are morally important in the same way that humans are? Deontologists who deny that claim can be understood as saying that a certain form of speciesism (or something close to it) is defensible, often because of differences in rationality between humans and other species. Nonetheless, deontologists in this tradition have argued that the respect we owe humans *generates* obligations toward animals. The reason for this is that humans share many features with other animals, such as our need for nourishment, drive to reproduce, and capacity for pain and pleasure. Because of that, a willingness to abuse or neglect the interests of animals can easily lead to, or perhaps itself even amount to, a failure to respect humans. Imagine seeing a small child gleefully stabbing a stuffed animal in its face with a toy knife. Even though no actual harm was being done, we might reasonably find the action intrinsically objectionable in the disregard it shows for suffering, and we might worry about what real harms it might lead to.

Consider the large-scale indifference to the needs of, e.g., wild terrestrial animals and finfish that are promoted by the technologies described in the chapter. That indifference is a failure to take seriously the death and suffering of large numbers of living creatures. Even if those creatures are less morally important than humans, that indifference resembles indifference toward large-scale human suffering, such as famines in third-world countries. Protecting the fragile respect we have, or ought to have, for other humans arguably then calls for cultivating more serious consideration of animals' needs.

The two lines of thought described in this commentary are compatible. One could hold *both* that animals deserve moral respect in their own right, *and* that we should regulate our attitudes toward them to support the respect we owe other humans. On either approach, deontology would, at a minimum, demand that developers of technology do not obscure the needs of animals. Fully meeting that minimal demand would require significant changes to existing platforms, but many smaller-scale responses are within reach.

African Ethics

By John Murungi

The call for a recognition of animal moral rights is voiced by a growing number of ethicists. However, the hurdle they seek to overcome has been created by those who have restricted rights to human beings. They see this restriction as species prejudice whereby human beings perceive themselves as exclusive holders of these rights. To them every being that is subject to suffering is entitled to inherent possession of these rights. Since animals suffer, they deserve to be included among the holders of these rights.

A comprehensive view of AI ought to consider the essence of technology. This may be obvious to many producers, owners, managers, and users of technology but what seems obvious may not be so. The question regarding the essence of technology remains fundamentally unasked and, to this extent, how AI stands in relation to technology remains unthought. Consequently, one of the dangers is that the essence of AI remains unthought. Thinking itself has become a matter of technology. Thinking has become technical. It is a matter of AI. AI and technology (its mother) have become increasingly the sole determiner of what thinking is. Co-optation of thinking by AI has increasingly become the reality of our time. AI has increasingly become dictatorial in determining the essence of thinking. The distinction between what is real and what is artificial has been so blurred that the real has become artificial and the artificial has become real. This affects the claims about the locality and the holders of moral rights.

Martin Heidegger, a major twentieth-century German philosopher makes an important contribution. He says,

> …. the essence of technology is by no means anything technological. Thus, we shall never experience our relationship to the essence of technology so long as we merely conceive and push forward the technological, put up with it, or evade it. Everywhere we remain unfree and chained to technology, whether we passionately affirm or deny it. But we are delivered ever to it in the worst when we regard it as something neutral, for this conception of it, to which today we particularly like to do homage, makes us utterly blind to the essence of technology.[56]

Heidegger reminds us that there is nothing technological about the essence of technology. Without going into details about his sense of the essence of technology, it seems that technology the logos of techno, is anthropology; that is, it is the logos of *anthropos*. Here, we should not have in view anthropology as a discipline taught in the academy. We should have in view a study of what it is to be a human being and what is at stake in human well-being.

Overwhelmingly, the view that ultimately technology is anthropology is rarely taken into consideration by those who engage in it. This is the case with those who present themselves or those who are presented as AI students or, more generally, as students of technology. They do not take this study as the study of who or what they are and what ultimately constitutes our well-being. They do not see themselves as children of AI or as mothers and fathers of AI.

Assuming that technology is anthropology, this anthropology remains radically problematic. In European modernity, Georg Wilhelm Friedrich Hegel, a prominent German philosopher, claimed that:

> The Negro, as already observed exhibits the natural man in his completely wild and untamed state. We lay aside all thought of reverence and morality — all that we call feeling — if we would rightly comprehend him; there is nothing harmonious with humanity to be found in this type of character.[57]

Hegel's words bring to mind Jean Paul Sartre, a French philosopher, words about the anti-Semite. Sartre argues that for the anti-Semite, the Jew that the anti-Semite takes to be quintessentially evil is an invention of his mind. If the Jew did not exist, he would have to invent one. The anti-Semite seeks to hide from himself that the Jew he has in mind is an invention of his mind, and he hides from this truth because accepting it would disclose his absolute dependence on the Jew for his being. Analogously, what Hegel takes as a Negro is nothing more than an invention of his own mind. In inventing the Negro, he invents himself. There are no Negroes except in the mind of Hegel and in the minds of those who think like him. Africans are Africans. They are not Negroes. To Africans, Negroes are not Africans.

It should be obvious that the realm of AI is a realm of invention. Apparently, this has become a cyberspace realm. The question of morality has become a cyberspace question. One aspect of this question is who morally governs in this realm and for whose benefit? Do animals have a say here? Here is where Hegel's version of humanism is relevant. Since he leaves out Africans from the community of human beings, it would appear that Africans have nothing to contribute to the moral governance of cyberspace or determining who or what has moral rights.

Let us bear in mind that cyberspace is a space that is absolutely cut off from terrestrial space. European modernity has invented its own version of colonialism. Could it be the case that cyberspace is a colonized space and is undergoing the process of colonization and that AI is the primary agent in this process?

It seems to me that what Africans can contribute to the discussion of morality in the cyberspace age is a liberatory ethics – an ethics that recognizes the oppressive weaponization of AI. The use of AI for the liberation of humankind is morally significant and it can open an avenue for animal liberation. In good conscience, we cannot recognize animals as bearers of moral rights if we inadequately have a sense of who or what we are.

One of the key concepts in African ethics is Ubuntu. A major claim in this ethics is the claim that we are therefore I am. The 'we' is more than an aggregate of 'Is.' It is a communal 'we.' In this ethics, what remains to be thought is who is included or excluded in it. The place of the animal is yet to be thought. To the extent it is not thought through, the human remains to be thought. Accordingly, the moral aspect of the 'we' remains to be thoroughly thought out. The governance of AI today is a major obstacle to the emergence of this thinking. The cyber anthropology – the Cyber Man stands in the way as a major obstacle.

Notes

1. Jeremy Bentham, *Introduction to the Principles of Morals and Legislation*, Chapter XVII §1.IV, note 1 (1789).
2. We say 'arguably' because one of the authors, Tse, is researching whether this barrier can be partially broken using new AI design methodologies, and new technologies. This research is still in its early stages, and has not yet produced results that would challenge the commonly-held view that animals cannot meaningfully participate in the design or feedback of machine learning processes.
3. R.A. Yusof, "Image Segmentation and Verification Based on Machine Learning for Vision Inspection of Chicken Slaughtering," *Journal of Physics: Conference Series*, 1447 (2019): 012024, doi: https://doi.org/10.1088/1742-6596/1447/1/012024.
4. Angus Taylor, "Philosophy and the Case for Animals," in *Canadian Perspectives on Animals and the Law*, ed. Peter Sankoff, Vaughan Black, and Katie Sykes, 9–30 (Toronto: Irwin Law, 2015), 16.
5. Katie Sykes, "Rethinking the Application of Canadian Criminal Law to Factory Farming," in *Canadian Perspectives on Animals and the Law*, ed. Peter Sankoff, Vaughan Black, and Katie Sykes, 33–56 (Toronto: Irwin Law, 2015). In this chapter, Sykes examines the law's inability to uphold animal welfare laws when it comes to factory farming.
6. Kelsey Piper, "'Ag-Gag Laws' Hide the Cruelty of Factory Farms from the Public. Courts Are Striking Them Down," *Vox*, January 11, 2019, https://www.vox.com/future-perfect/2019/1/11/18176551/ag-gag-laws-factory-farms-explained.
7. Marc Bekoff, *Encyclopedia of Animal Rights and Animal Welfare* (Santa Barbara, Calif, 2009).
8. Sabine Sczesny, Magda Formanowicz, and Franziska Moser, "Can Gender-Fair Language Reduce Gender Stereotyping and Discrimination?" *Frontiers in Psychology*, 7 (2016), doi:https://www.frontiersin.org/article/10.3389/fpsyg.2016.00025.
9. John R. Rickford, *Raciolinguistics: How Language Shapes Our Ideas About Race* (Oxford, New York: Oxford University Press, 2016).
10. Tracey L. Gendron, E. Ayn Welleford, Jennifer Inker, and John T. White, "The Language of Ageism: Why We Need to Use Words Carefully," *The Gerontologist*, 56, no. 6 (December 1, 2016): 997–1006, doi: https://doi.org/10.1093/geront/gnv066.
11. Arron Stibbe, "Language, Power and the Social Construction of Animals," *Society & Animals*, 9, no. 2 (2001): 145–161; Joan Dunayer, "Review of Animal Equality: Language and Liberation," *GEMA Online Journal of Language Studies*, 7, no. 2 (2007): 1.

12. Jonas R. Kunst, and Sigrid M. Hohle, "Meat Eaters by Dissociation: How We Present, Prepare and Talk about Meat Increases Willingness to Eat Meat by Reducing Empathy and Disgust," *Appetite,* 105 (2016): 758–74, doi: https://doi.org/10.1016/j.appet.2016.07.009.
13. Nikola Mrkšić, Diarmuid Ó Séaghdha, Blaise Thomson, Milica Gašić, Lina M. Rojas-Barahona, Pei-Hao Su, David Vandyke, Tsung-Hsien Wen, and Steve Young, "Counter-Fitting Word Vectors to Linguistic Constraints," Proceedings of the 2016 Conference of the North American Chapter of the Association for Computational Linguistics: Human Language Technologies, 142–48, San Diego, California: Association for Computational Linguistics, 2016, doi: https://doi.org/10.18653/v1/N16-1018; Ivan Vulić, and Nikola Mrkšić, "Specialising Word Vectors for Lexical Entailment," ArXiv:1710.06371 [Cs], October 17, 2017, http://arxiv.org/abs/1710.06371.
14. Arguably, the species naming itself is not speciesist, but it was the historical events that the name of the species recorded that was speciesist, because this species likely emerged because humans wanted to domesticate wild fowls to be suitable to be trained for cockfighting and raised to become food – there were simply no "chickens" before humans chose to domesticate wild fowls.
15. George A. Miller and Christiane Fellbaum, "WordNet Then and Now," *Language Resources and Evaluation,* 41, no. 2 (2007): 209–14.
16. Solaiman, Irene, and Christy Dennison, "Process for Adapting Language Models to Society (PALMS) with Values-Targeted Datasets," *Advances in Neural Information Processing Systems,* 34:5861–73, Curran Associates, Inc., 2021, https://proceedings.neurips.cc/paper/2021/hash/2e855f9489df0712b4bd8ea9e2848c5a-Abstract.html.
17. Jahna Otterbacher, Alessandro Checco, Gianluca Demartini, and Paul Clough, "Investigating User Perception of Gender Bias in Image Search: The Role of Sexism," *The 41st International ACM SIGIR Conference on Research & Development in Information Retrieval,* 933–36, SIGIR '18, New York, NY, USA: Association for Computing Machinery, 2018, https://doi.org/10.1145/3209978.3210094; Megan Garcia, "Racist In the Machine: The Disturbing Implications Of Algorithmic Bias," *World Policy Journal,* 33, no. 4 (2016): 111–17; Safiya Umoja Noble, *Algorithms of Oppression: How Search Engines Reinforce Racism* (New York University Press, 2018), doi: https://doi.org/10.18574/9781479833641.
18. Robert Epstein, and Ronald E. Robertson, "The Search Engine Manipulation Effect (SEME) and Its Possible Impact on the Outcomes of Elections," *Proceedings of the National Academy of Sciences,* 112, no. 33 (2015): E4512–21, doi: https://doi.org/10.1073/pnas.1419828112.
19. Justyna Stypinska, "AI Ageism: New Forms of Age Biases and Age Discrimination in the Era of Algorithms and Artificial Intelligence," *Research Gate* (2021), accessed July 1, 2022, https://www.researchgate.net/profile/Justyna-Stypinska/publication/354438029_Title_AI_ageism_new_forms_of_age_biases_and_age_discrimination_in_the_era_of_algorithms_and_artificial_intelligence/links/61387fb4c76de21e319c9e9d/Title-AI-ageism-new-forms-of-age-biases-and-age-discrimination-in-the-era-of-algorithms-and-artificial-intelligence.pdf.
20. Liwen Vaughan, and Mike Thelwall, "Search Engine Coverage Bias: Evidence and Possible Causes," *Information Processing & Management* 40, no. 4 (2003): 693–707, doi: https://doi.org/10.1016/S0306-4573(03)00063-3.
21. Google webpage search seems to be much more affected by IP locations, so we did the research with image search. An additional reason is that, speculatively, image search might have higher impact on young children than webpage search when it comes to learning about animals.
22. This avoids the Chrome browser, which one of the authors had previously used.
23. Toby G. Knowles, Steve C. Kestin, Susan M. Haslam, Steven N. Brown, Laura E. Green, Andrew Butterworth, Stuart J. Pope, Dirk Pfeiffer, and Christine J. Nicol, "Leg Disorders in Broiler Chickens: Prevalence, Risk Factors and Prevention," *PLOS One,* 2, (2008): e1545, doi: https://doi.org/10.1371/journal.pone.0001545.

24. People's Dispensary for Sick Animals (PDSA), "How Many Pets Are There in the UK?" *People's Dispensary for Sick Animals* (Website), 2022, accessed May 5, 2022, "https://www.pdsa.org.uk/what-we-do/pdsa-animal-wellbeing-report/uk-pet-populations-of-dogs-cats-and-rabbits.

25. Royal Society for the Protection of Animals (RSPCA), "Facts and Figures," *Royal Society for the Protection of Animals* (Website), accessed July 1, 2022, https://www.pdsa.org.uk/what-we-do/pdsa-animal-wellbeing-report/uk-pet-populations-of-dogs-cats-and-rabbits.

26. Which means that this figure does not count aquatic mammals, reptiles, amphibians, and invertebrates.

27. Animal Equality, "The Deadly Fish Industry," *Animal Equality* (Website), 2022, accessed July 1, 2022, https://animalequality.org/issues/fish/; Kathy Hessler, Rebecca Jenkins, and Kelly Levenda, "Cruelty to Human and Nonhuman Animals in the Wild-Caught Fishing Industry," *Sustainable Development Law & Policy,* 18, no. 1 (2017): 30; Helena Horton, "No Routine Checkups on Welfare of Fish at Slaughter, Officials Admit," *The Guardian,* November 23, 2021, accessed July 1, 2022, https://www.theguardian.com/environment/2021/nov/23/no-routine-checkups-on-welfare-of-fish-at-slaughter-officials-admit.

28. Robyn J. Crook, "Behavioral and Neurophysiological Evidence Suggests Affective Pain Experience in Octopus," *Iscience,* 24, no. 3 (March 19, 2021): 102229, doi: https://doi.org/10.1016/j.isci.2021.102229; Robert W. Elwood, "Pain and Suffering in Invertebrates?" *ILAR Journal,* 52, no. 2 (2011): 175–84, doi: https://doi.org/10.1093/ilar.52.2.175.

29. Department for Environment, Food & Rural Affairs (UK), "Lobsters, Octopus and Crabs Recognised as Sentient Beings," *UK Government* (Press Release), November 19, 2021, https://www.gov.uk/government/news/lobsters-octopus-and-crabs-recognised-as-sentient-beings.

30. One of the authors is a Chinese speaker.

31. Sentient Media Editorial Team, "Farmed Animals: The Exploitation of Animals on Factory Farms," *Sentient Media,* May 7, 2020, https://sentientmedia.org/farmed-animals/.

32. We also did a search for "Concentrated Animal Feeding Operations," which returned similar results to "factory farm," but the operations shown appear to be less filthy and crowded on average.

33. American Society for the Protection of Animals (ASPCA), "Protecting Farm Animals," *American Society for the Protection of Animals* (Website), 2022, accessed July 1, 2022, https://www.aspca.org/protecting-farm-animals.

34. Aquabyte, "Machine Learning, Technology, and Sustainability: A Partner. More than Just a Camera," *Aquabyte* (Website), 2022, accessed July 1, 2022, https://aquabyte.ai/.

35. Ron Tardiff, "The Current State of Sea Lice Management," *Aquaculture Stewardship Council Foundation,* April 25, 2019, https://www.asc-aqua.org/the-current-state-of-sea-lice-management/.

36. Liza Mayer, "In Norway, Counting Sea Lice Manually May Be on Its Way Out," *Aquaculture North America,* February 11, 2021, https://www.aquaculturenorthamerica.com/in-norway-counting-sea-lice-manually-may-be-on-its-way-out/.

37. Aquabyte, "Machine Learning."; Rodrigo García, Jose Aguilar, Mauricio Toro, Angel Pinto, and Paul Rodríguez, "A Systematic Literature Review on the Use of Machine Learning in Precision Livestock Farming," *Computers & Electronics in Agriculture,* 179 (2020): 105826, doi: https://doi.org/10.1016/j.compag.2020.105826.

38. Umitron, "Create Computerized Aquaculture," *Umitron* (Website), 2019, accessed July 1, 2022, https://umitron.com/en/service.html.

39. Elodie F. Briefer, Ciara C.R. Sypherd, Pavel Linhart, Lisette M.C. Leliveld, Monica Padilla de la Torre, Eva R. Read, Carole Guérin, *et al,* "Classification of Pig Calls Produced from Birth to Slaughter According to Their Emotional Valence and Context

of Production," *Scientific Reports,* 12, no. 1 (2022): 3409, doi: https://doi.org/10.1038/s41598-022-07174-8; Satyabrata Aich, Sabyasachi Chakraborty, Jong-Seong Sim, Dong-Jin Jang, and Hee-Cheol Kim, "The Design of an Automated System for the Analysis of the Activity and Emotional Patterns of Dogs with Wearable Sensors Using Machine Learning," *Applied Sciences,* 9, no. 22 (2019): 4938, doi: https://doi.org/10.3390/app9224938.

40. See: Thomas Nagel, "What is it Like to Be a Bat," *Philosophical Review,* 83, no. 4 (1974): 435–450, doi: https://doi.org/10.2307/2183914, in which he explores the 'mind-body problem' and the limits of our ability to imagine the subjective phenomenal experience of other living beings (and sometimes even ourselves).
41. Genesis 1:28.
42. Numbers 22:32; Guide for the Perplexed 3:17.
43. Exodus 23:5.
44. Deuteronomy 25:4.
45. Exodus 20:10; Deuteronomy 5:14.
46. Rashi, Exodus 23:12.
47. *Shabbat* 128b.
48. Genesis 1:29; Rashi, ad.loc.
49. Genesis 9:3–4.
50. Malbim, Genesis 9:3.
51. Genesis 9:5–6.
52. Guide for the Perplexed 3:26.
53. Psalms 145:9.
54. *Bava Metziah* 85a.
55. Jonathan Safron Foer, *Eating Animals* (New York: Little Brown & Company, 2009), 197–198.
56. Martin Heidegger, *The Question Concerning Technology and Other Essays,* trans. William Lovett (New York and London, Garland Publishing, 1977), 4.
57. Georg W.F. Hegel, *The Philosophy of History,* trans. J. Sibree, intro. C.J. Friedrich (New York: Dover Publications, 1956), 93.

Epilogue

This book is a journey through case studies in data science that raises ethical dilemmas and the way they should be analyzed from a range of ethical perspectives. Rather than trying to provide the reader with ready-made answers to similar situations, we have instead portrayed a framework that will allow a wide, diverse, and foundational starting point in addressing such ethical dilemmas.

The first step in such cases is always to actually realize that an ethical dilemma is existing and needs to be resolved. With the advance of technology and the blurring of the borders between the virtual and the real, this first step is many times the most challenging one and the identification and understanding of the ethical dilemma is both the most difficult and important part of the task.

We hope that the cases brought in this book will assist the reader in identifying such issues going forward and will sharpen the reader's observation ability in detecting such issues. The second step is addressing these ethical dilemmas, preferably from more than one ethical approach. We provided the foundations and the application of the ethical approaches brought in this book, with regard to the cases at hand.

Synthesizing, analyzing, and internalizing these two steps should provide the data scientist the sufficient background to conduct their work in a way that will be as ethical as possible and, hopefully, to develop a good deal of wisdom along the way.

Index

Printed in the United States
by Baker & Taylor Publisher Services